Ben Stacy Jerrik (Ed.)

MS Stena Cambria

Ben Stacy Jerrik (Ed.)

MS Stena Cambria

Sealink, Ventouris Ferries, Mediterranean Sea, English Channel

Part Press

Imprint

Publisher:
Part Press is a trademark of
International Book Market Service Ltd., 17 Rue Meldrum, Beau Bassin, 1713-01 Mauritius
Email: info@bookmarketservice.com
Website: www.bookmarketservice.com

Published in 2012

Printed in: U.S.A., U.K., Germany. This book was not produced in Mauritius.

ISBN: 978-620-1-66225-4

Contents

Articles

References

The *Isla de Botafoc* in 2007

Career	
Name:	*Bari* (2010 – present) *Winner 9* (2010) *Isla de Botafoc* (1999-2010) *Stena Cambria* (1991-1999) *St Anselm* (1980-1991)[1]
Operator:	Ventouris Ferries (2010 – present) Balearia (2003-2010) UMAFISA (1999-2003) P&O Stena Line (1998-1999) Stena Line (1991-1998) Sealink (1980-1991)[1]
Port of registry:	Limassol, Cyprus [1]
Builder:	Harland & Wolff, Belfast
Yard number:	1715
Launched:	5 December 1979 [2]
Christened:	4 December 1979 [2]
Maiden voyage:	27 October 1980 [3]
Identification:	IMO number: 7813937 [1]
General characteristics	
Tonnage:	7,003 gross register tons (GRT) [1]
Length:	129.65 m (**unknown operator: u'strong'** ft) [1]
Beam:	21.62 m (**unknown operator: u'strong'** ft) [1]
Draught:	4.84 m (**unknown operator: u'strong'** ft) [1]
Propulsion:	Pielstick [1]
Speed:	18 kn (**unknown operator: u'strong'** km/h) [1]

Capacity:	1,200 passengers [1]

MS *Stena Cambria* is a ferry launched in 1980 as the *St Anselm* [4] for Sealink. Starting life on the Dover-Calais, she now operates as the *Bari* with Ventouris Ferries in the Mediterranean.

Sealink

The *Stena Cambria* started life as the *St Anselm*, order by and for Sealink services in the English Channel [3] . She was launched at the Harland & Wolff shipyard in Belfast [3] on 5 December 1979 and completed her maiden voyage on 27 October 1980 [3] .

After two years in service, on 31 December 1982 [3] , the *St Anselm* returned to Belfast for a £750,000 extension to her aft decks [3] . This extension provided an enlarged duty free shopping area and additional accommodation, increasing her tonnage by 402 GT (gross tonnage) to 7,405 GT (gross tonnage) and her passenger capacity to 1,400 [3] . On 28 March 1983 during her trip back to Dover through the Irish Sea, the *St Anselm* was temporarily brought into service on the Fishguard-Rosslare route following the failure of *Stena Normandica*'s engines [3] .

In 1990, the *St Anselm* was displaced from the Dover-Calais route by Sealink's acquisition of the *Fantasia*, the *St Anselm* instead took up a service operating between Folkestone and Boulogne [3] . Not long after this transition, Sealink was acquired by Stena Line, in 1991 the *St Anselm* was transferred to Holyhead [3] , being renamed the *Stena Cambria* as a refit relief during February and March [3] , briefly returning to Dover before resuming service in the Irish Sea in July that same year [3] .

Stena Line

In July 1995, SNAT announced they would be terminating the pooling agreement that had been in place with Stena Sealink Line, which as a result, dropped the "Sealink" name, becoming just Stena Line. As a result of the loss of the two French vessels, Stena Line transferred the *Stena Cambria* and the *Stena Empereur* back to the Dover-Calais route for the 1996 season [5] . In March 1997 she provided refit relief in Stranraer and on 3 May 1997 became the last conventional ferry to sail out of Holyhead having provided overhaul relief for the *Stena Explorer* [3] . In March 1998 she transferred to the Newhaven-Dieppe route under the newly formed P&O Stena Line, staying there until the routes withdrawal in January 1999 [5] .

Post P&O Stena Line

Following the routes withdrawal in January 1999, the *Stena Cambria* was laid up in Zeebrugge awaiting sale [5] , eventually being purchased by UMAFISA, entering service for them between Barcelona and Ibiza that November [5] as the *Isla de Botafoc*. In August 2003, UMAFISA and the *Isla de Botafoc* were acquired by rivals, Balearia [5] . She continued services with Balearia to Ibiza and Menorca [5] until her eventual sale to Ventouris Ferries of Greece in 2010, briefly being renamed *Winner 9* and then to the *Bari* [1] .

Sister ships

The *Stena Cambria* has three sister ships;

- *Stena Caledonia*
- *Stena Antrim*
- *Stena Galloway*

References

[1] "M/F Bari" (http://www.ferry-site.dk/ferry.php?id=7813937&lang=en). Ferry-site.dk. . Retrieved 28 June 2010.

[2] "MS *Stena Cambria* Part One" (http://www.hhvferry.com/stan1.html). HHVFerry.com. . Retrieved 28 June 2010.

[3] "*Stena Cambria*" (http://www.sealink-holyhead.com/sealink/ships/cambria/home.html). Sealink-Holyhead.com. . Retrieved 28 June 2010.

[4] "*Stena Cambria*" (http://www.simplonpc.co.uk/PandO-StenaLine.html#anchor79200). SimplonPC.co.uk. . Retrieved 28 June 2010.

[5] "MS *Stena Cambria* Part Two" (http://www.hhvferry.com/stan2.html). HHVFerry.com. . Retrieved 28 June 2010.

Sealink

Sealink	
Industry	Maritime
Fate	Privatised 1984, resold 1991 (becoming Stena Line (UK))
Headquarters	Great Britain and adjacent waters
Products	Sea transport
Parent	British Transport Commission (until 1962), British Railways Board (since 1962)

Sealink was a ferry company based in the United Kingdom from 1970 to 1984, operating services to France, Belgium, the Netherlands, Channel Islands, Isle of Wight and Ireland.

Ports served by the company included: Dover, Folkestone, Newhaven, Southampton and Harwich for services to the European continent; Holyhead, Fishguard, Heysham and Stranraer for services to Ireland and the Isle of Man; services to the Channel Islands from Weymouth & Portsmouth. The Isle of Wight was also served from Portsmouth and Lymington.

Sealink also operated the "Steamer" Passenger Ferry services on Windermere in Cumbria until privatisation when these were passed to the newly reformed Windermere Iron Steamboat Company (now Windermere Lake Cruises Ltd).

History

Sealink was originally the brand name for the ferry services of British Rail which ran shipping services in the UK and Ireland. Services to France, Belgium and the Netherlands were also run by Sealink UK as part of the Sealink consortium which also used ferries owned by French national railways, the *SNCF*, the Belgian Maritime Transport Authority, *Regie voor Maritiem Transport / Regie des transports maritimes (RMT/RTM)* and the Dutch Stoomvaart Maatschappij Zeeland (Zeeland Steamship Company).

Historically, the shipping services were exclusively an "extension" of the railways across the English Channel and the Irish Sea in order to provide through, integrated services to Europe and Ireland. As international travel became more popular in the late 1960s and before air travel became generally affordable, the responsibility for shipping services was taken away from the British Rail Regions and in 1969 centralised in a new division - British Rail Shipping and International Services Division.

With the advent of car ferry services the old passenger-only ferries were gradually replaced by roll-on-roll-off ships catering both for motorists and rail passengers as well as road freight. However, given that there was now competition in the form of other ferry companies offering crossings to motorists, it became necessary to market the services in a normal business fashion (as opposed to the previous almost monopolistic situation). Thus, with the other partners mentioned above, the brand name Sealink was introduced for the consortium.

As demand for international rail travel declined and the shipping business became almost exclusively dependent on passenger and freight vehicle traffic, the ferry business as was incorporated as **Sealink UK Limited** in 1978, a wholly owned subsidiary of the British Railways Board (BRB), but still as part of the Sealink consortium. In 1979 Sealink acquired Manx Line offering services to the Isle of Man from Heysham.

On 27 July 1984 the UK Government sold Sealink UK Ltd to Bermuda-based Sea Containers for £66m. The company was renamed **Sealink British Ferries**. 7 years later, In 1991 Sea Containers sold Sealink British Ferries to Stena Line. The sale excluded the operations of Hoverspeed, the Isle of Wight services and the share in the Isle of

Man Steam Packet Company, as well as the Port of Heysham. The new Swedish owners changed the company name to **Sealink Stena Line** then again a few years later to **Stena Sealink Line**.

In 1996 the Sealink name disappeared when the UK services were re-branded as Stena Line. The agreement with the SNCF on the Dover-Calais route also ended at this time and the French run Sealink services were rebranded as SeaFrance.

Livery

The livery from 1984 to 1995 was a distinctive blue-on-white. Previously, the British Rail logo had been used, with a BR corporate monastral blue hull, white upperworks and black-topped red funnel. From 1972/73 "Sealink" was displayed in white Rail Alphabet typeface on the side of the hull. Prior to 1964/65 the colours were black hull, white upperworks and black-topped buff funnel.

Sealink house flag

An inverted version of the BR symbol was used on Sealink's funnels and flags.[1] This was because:

1. The inverted arrows make an "S" for Sealink when in that direction.
2. While railways in the UK pass right to right (i.e. drive on the left), ships have to pass left to left (i.e. drive on the right), so this version was more accurate.
3. On the ships' funnels the symbol was reversed on the port side but correct on the starboard side, so the 'top' arrow was always pointing towards the bow of the ship.

British Rail owned ships had red funnels with a white logo. Elsewhere in the company the symbol was white on blue.

Hovercraft

In the 1960s, British Rail started hovercraft services from Dover to Calais and Boulogne, and also across the Solent to the Isle of Wight. Rather than use the name Sealink, the services were marketed as "Seaspeed" and 'Solent Seaspeed'. Seaspeed merged with rival Hoverlloyd in 1981 to create Hoverspeed.

References

[1] British Rail (http://www.fotw.net/flags/gb_brail.html)

External links

- stenaline.com (http://www.stenaline.com)
- Sealink-Holyhead.com (http://www.sealink-holyhead.com) a guide to the history of the sea route between Holyhead and Dún Laoghaire
- Kangaroo Island Sealink (http://www.sealink.com.au) Ferry operator between Australian mainland and Kangaroo Island (South Australia)
- Picture of the Sealink car ferry the *MV Hengist* beached at Folkestone during the [[Great Storm of 1987 (http://www.webcitation.org/query?url=http://www.geocities.com/treiziste/heng_beached.jpg&date=2009-10-25+23:16:42)] showing the blue-and-white livery.]
- Express Apollon,ex Senlac,3d video (http://www.youtube.com/watch?v=jlWjjxNRCXY)

Ventouris_Ferries

Founded	1978
Headquarters	Piraeus, Greece
Area served	Adriatic Sea
Services	Passenger transportation, Freight transportation
Website	www.ventouris.gr [1]

Ventouris Ferries is a Greek company that operates ferry services from Italy to Greece and Albania.

Routes

Ventouris Ferries operates two routes across the Adriatic Sea.

- Bari - Durres
- Bari - Corfu - Igoumenitsa

Fleet

Ventouris Ferries currently (June 2010) operates a fleet of four RORO ferries.[2]

Ship	Built	Entered service	Route	Tonnage	Flag	Notes
MS *Bari*	1980	2010	Bari - Durres	7,197 GRT	Cyprus	
MS Polaris	1975	1991	Bari - Corfu - Igoumenitsa	20,326 GRT	Cyprus	
MS Rigel	1973	2007	Bari - Durres	12,281 GRT	Panama	
MS Seatrade	1973	2008	Bari - Corfu - Igoumenitsa	17,991 GRT	Cyprus	

References

[1] http://www.ventouris.gr
[2] http://www.faktaomfartyg.nu/ventouris_ferries.htm

English_Channel

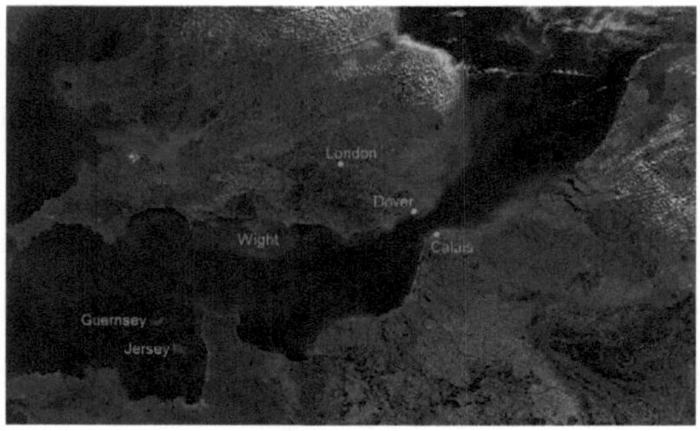

Satellite view of the English Channel

The **English Channel** (French: *la Manche*, Breton: *Mor Breizh*, Cornish: *Mor Bretannek*), often referred to simply as the **Channel**, is an arm of the Atlantic Ocean that separates southern England from northern France, and joins the North Sea to the Atlantic. It is about 560 km (**unknown operator: u'strong'** mi) long and varies in width from 240 km (**unknown operator: u'strong'** mi) at its widest to 34 km (**unknown operator: u'strong'** mi) in the Strait of Dover.[1] It is the smallest of the shallow seas around the continental shelf of Europe, covering an area of some 75000 km^2 (**unknown operator: u'strong'** sq mi).[2]

Geography

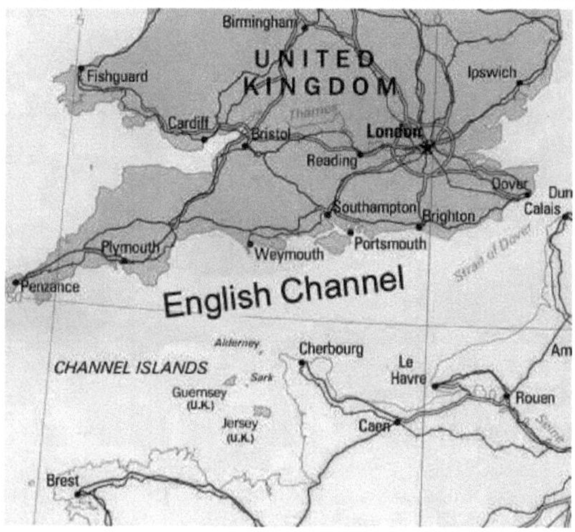

Map of the English Channel

The International Hydrographic Organization defines the limits of the English Channel as follows:[3]

> *On the West.* A line joining Isle Vierge (48°38′23″N 4°34′13″W) to Lands End (50°04′N 5°43′W).
>
> *On the East.* The Southwestern limit of the North Sea. The IHO defines the southwestern limit of the North Sea as "a line joining the Walde Lighthouse (France, 1°55'E) and Leathercoat Point (England, 51°10'N)".[3] The Walde Lighthouse is 6 km east of Calais (50°59′06″N 1°55′00″E), and Leathercoat Point is at the north end of St Margaret's Bay, Kent (51°10′00″N 1°24′00″E). The Strait of Dover (French: *Pas de Calais*), at the Channel's eastern end is its narrowest point, while its widest point lies between Lyme Bay and the Gulf of Saint Malo near its midpoint.[1] It is relatively shallow, with an average depth of about 120 m (**unknown operator: u'strong'** ft) at its widest part, reducing to a depth of about 45 m (**unknown operator: u'strong'** ft) between Dover and Calais. From there eastwards the adjoining North Sea continues to shallow to about 26 m (**unknown operator: u'strong'** ft) in the Broad Fourteens where it lies over the watershed of the former land bridge between East Anglia and the Low Countries. It reaches a maximum depth of 180 m (**unknown operator: u'strong'** ft) in the submerged valley of Hurds Deep, 30 mi (**unknown operator: u'strong'** km)

west-northwest of Guernsey.[4] The eastern region along the French coast between Cherbourg and the mouth of the Seine river at Le Havre is frequently referred to as the *Bay of the Seine (French: Baie de Seine).*[5] There are several major islands in the Channel, the most notable being the Isle of Wight off the English coast, and the Channel Islands, British Crown Dependencies off the coast of France. The Isles of Scilly off the far southwest coast of England are not generally counted as being in the Channel. The coastline, particularly on the French shore, is deeply indented; several small islands close to the coastline, including Chausey and Mont Saint-Michel, are within French jurisdiction. The Cotentin Peninsula in France juts out into the Channel, and the Isle of Wight creates a small parallel channel known as the Solent in English waters. The Celtic Sea is to the west of the Channel. The Channel is of geologically recent origins, having been dry land for most of the Pleistocene period. It is thought to have been created between 450,000 and 180,000 years ago by two catastrophic glacial lake outburst floods caused by the breaching of the Weald–Artois anticline, a ridge that held back a large proglacial lake in the Doggerland region, now submerged under the North Sea. The flood would have lasted for several months, releasing as much as one million cubic metres of water per second. The cause of the breach is not known but may have been an earthquake or the build-up of water pressure in the lake. The flood carved a large bedrock-floored valley down the length of the Channel, leaving behind streamlined islands and longitudinal erosional grooves characteristic of catastrophic megaflood events.[6] [7] It destroyed the isthmus that connected Britain to continental Europe, although a land bridge across the southern North Sea would have existed intermittently at later times after periods of glaciation resulted in lower sea levels.[8] For the UK Shipping Forecast the Channel is divided into the following areas, from the west:

- Plymouth
- Portland
- Wight
- Dover

Name

The name *English Channel* has been widely used since the early 18th century, possibly originating from the designation *Engelse Kanaal* in Dutch sea maps from the 16th century onwards. Historically, it was also been known as the *British Channel*.[9] [10] and as the *British Sea* and was called the *Oceanus Britannicus* by the 2nd-century geographer Ptolemy. The same name is used on an Italian map of about 1450, which gives the alternative name of *canalites Anglie*—possibly the first recorded use of the *Channel* designation.[11]

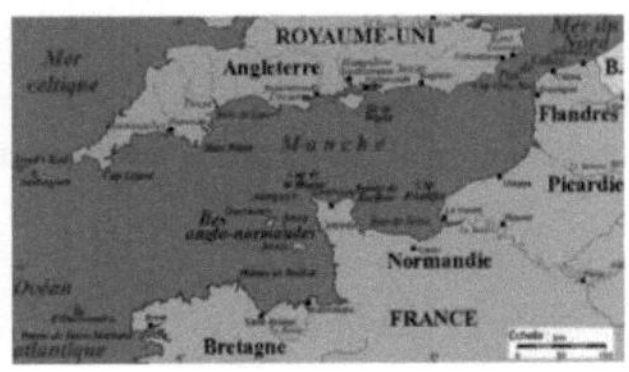

Map with French nomenclature

The French name *la Manche* has been in use since at least the 17th century.[2] The name is usually said to refer to the Channel's sleeve (French: *le manche*) shape. However, it is sometimes claimed to derive from a Celtic word meaning *channel* that is also the source of the name for the Minch in Scotland.[12]

In Spain and most Spanish speaking countries the Channel is referred to as *el Canal de la Mancha*. In Portuguese it is known as *Canal da Mancha*. This is not a translation from French: in Portuguese and Spanish, *mancha* means *stain*, while the word for sleeve is *manga* - which prompts an early phonetic bad translation from French. Other languages also use this name, such as Greek (*Κανάλι της Μάγχης*) and Italian (*la Manica*). The German name is *Ärmelkanal*, literally *sleeve-channel*.

History

Before the end of the Devensian glaciation (the most recent ice age) around 10,000 years ago, the British Isles were part of continental Europe. During this period the North Sea and almost all of the British Isles were covered with ice. The sea level was about 120 m (**unknown operator: u'strong'** ft) lower than it is today, and the channel was an expanse of low-lying tundra, through which passed a river which drained the Rhine and Thames towards the Atlantic to the west. As the ice sheet melted, a large freshwater lake formed in the southern part of what is now the North Sea. As the meltwater could still not escape to the north (as the northern North Sea was still frozen) the outflow channel from the lake entered the Atlantic Ocean in the region of Dover and Calais.

This precious stone set in the silver sea,
Which serves it in the office of a wall
Or as a moat defensive to a house,
Against the envy of less happier lands.

William Shakespeare, *Richard II* (Act II, Scene 1)

The Channel has been the key natural defence for Britain, halting invading armies while in conjunction with control of the North Sea allowing Britain to blockade the continent. The most significant failed invasion threats came when the Dutch and Belgian ports were held by a major continental power, e.g. from the Spanish Armada in 1588, Napoleon during the Napoleonic Wars, and Nazi Germany during World War II. Successful invasions include the Roman conquest of Britain and the Norman Conquest in 1066, while the concentration of excellent harbours in the Western Channel on Britain's south coast made possible the largest invasion of all time, the Normandy landings in 1944. Channel naval battles include the Battle of Goodwin Sands (1652), the Battle of Portland (1653), the Battle of La Hougue (1692) and the engagement between USS *Kearsarge* and CSS *Alabama* (1864).

In more peaceful times the Channel served as a link joining shared cultures and political structures, particularly the huge Angevin Empire from 1135 to 1217. For nearly a thousand years, the Channel also provided a link between the Modern Celtic regions and languages of Cornwall and Brittany. Brittany was founded by Britons who fled Cornwall and Devon after Anglo-Saxon encroachment. In Brittany, there is a region known as "Cornouaille" (Cornwall) in French and "Kernev" in Breton[13] In ancient times there was also a "Domnonia" (Devon) in Brittany as well.

In February 1684 (New style), ice formed on the sea in a belt 3 miles (**unknown operator: u'strong'** km) wide off the coast of Kent and 2 miles (**unknown operator: u'strong'** km) wide on the French side.[14] [15]

Route to the British Isles

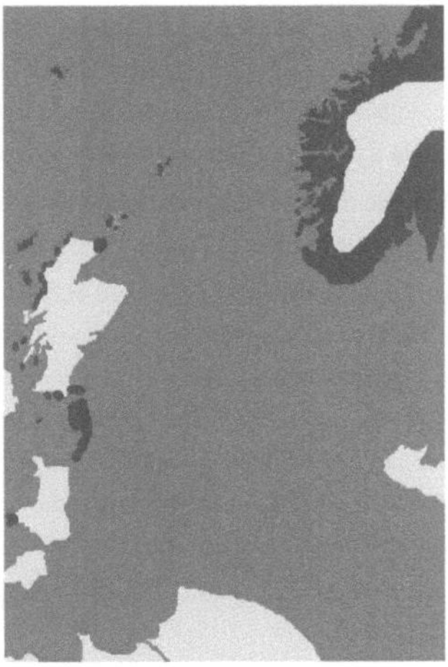

The approximate extent of Old Norse and related languages in the early 10th century around the North Sea. The red area is the distribution of the dialect Old West Norse, the orange area Old East Norse, and the green area the other Germanic languages with which Old Norse still retained some mutual intelligibility

Diodorus Siculus and Pliny[16] both suggest trade between the rebel celtic tribes of Armorica and Iron Age Britain flourished. In 55 BC Julius Caesar invaded, claiming that the Britons had aided the Veneti against him the previous year. He was more successful in 54 BC, but Britain was not fully established as part of the Roman Empire until completion of the invasion by Aulus Plautius in 43 AD. A brisk and regular trade began between ports in Roman Gaul and those in Britain. This traffic continued until the end of Roman rule in Britain in 410 AD, after which the early Anglo-Saxons left less clear historical records.

In the power vacuum left by the retreating Romans, the Germanic Angles, Saxons, and Jutes began the next great migration across the North Sea. Having already been used as mercenaries in Britain by the Romans, many people from these tribes crossed during the Migration Period, conquering and perhaps displacing the native Celtic populations.[17]

Norsemen and Normans

The Hermitage of St Helier lies in the bay off St. Helier and is accessible on foot at low tide

The attack on Lindisfarne in 793 is generally considered the beginning of the Viking Age. For the next 250 years the Scandinavian raiders of Norway, Sweden, and Denmark dominated the North Sea, raiding monasteries, homes, and towns along the coast and along the rivers that ran inland. According to the Anglo-Saxon Chronicle they began to settle in Britain in 851. They continued to settle in the British Isles and the continent until around 1050.[18]

The fiefdom of Normandy was created for the Viking leader Rollo (also known as Robert of Normandy). Rollo had besieged Paris but in 911 entered vassalage to the king of the West Franks Charles the Simple through the Treaty of St.-Claire-sur-Epte. In exchange for his homage and fealty, Rollo legally gained the territory he and his Viking allies had previously conquered. The name "Normandy" reflects Rollo's Viking (i.e. "Northman") origins.

The descendants of Rollo and his followers adopted the local Gallo-Romantic language and intermarried with the area's inhabitants and became the Normans – a Norman French-speaking mixture of Scandinavians, Hiberno-Norse, Orcadians, Anglo-Danish, and indigenous Franks and Gauls.

Rollo's descendant William, Duke of Normandy became king of England in 1066 in the Norman Conquest culminating at the Battle of Hastings, while retaining the fiefdom of Normandy for himself and his descendants. In 1204, during the reign of King John, mainland Normandy was taken from England by France under Philip II, while insular Normandy (the Channel Islands) remained under English control. In 1259, Henry III of England recognised the legality of French possession of mainland Normandy under the Treaty of Paris. His successors, however, often fought to regain control of mainland Normandy.

With the rise of William the Conqueror the North Sea and Channel began to lose some of their importance. The new order oriented most of England and Scandinavia's trade south, toward the Mediterranean and the Orient.

Although the British surrendered claims to mainland Normandy and other French possessions in 1801, the monarch of the United Kingdom retains the title Duke of Normandy in respect to the Channel Islands. The Channel Islands (except for Chausey) are a Crown dependency of the British Crown. Thus the Loyal Toast in the Channel Islands is *La Reine, notre Duc* ("The Queen, our Duke"). The British monarch is understood to *not* be the Duke of Normandy in regards of the French region of Normandy described herein, by virtue of the Treaty of Paris of 1259, the surrender of French possessions in 1801, and the belief that the rights of succession to that title are subject to Salic Law which excludes inheritance through female heirs.

French Normandy was occupied by English forces during the Hundred Years' War in 1346–1360 and again in 1415–1450.

England and Britain: The naval superpowers

From the reign of Elizabeth I, English foreign policy concentrated on preventing invasion across the Channel by ensuring no major European power controlled the potential Dutch and Flemish invasion ports. Her climb to the pre-eminent sea power of the world began in 1588 as the attempted invasion of the Spanish Armada was defeated by the combination of outstanding naval tactics by the English under command of Charles Howard, 1st Earl of Nottingham with Sir Francis Drake second in command, and the following stormy weather. Over the centuries the Royal Navy slowly grew to be the most powerful in the world.[19]

The building of the British Empire was possible only because the Royal Navy exercised unquestioned control over the seas around Europe, especially the Channel and the North Sea. During the Seven Years' War, France attempted to launch an invasion of Britain. To achieve this France needed to gain control of the Channel for several weeks, but was thwarted following the British naval victory at the Battle of Quiberon Bay in 1759.

Another significant challenge to British domination of the seas came during the Napoleonic Wars. The Battle of Trafalgar took place off the coast of Spain against a combined French and Spanish fleet and was won by Admiral Horatio Nelson, ending Napoleon's plans for a cross-Channel invasion and securing British dominance of the seas for over a century.

First World War

The exceptional strategic importance of the Channel as a tool for blockade was recognised by the First Sea Lord Admiral Fisher in the years before World War I. "Five keys lock up the world! Singapore, the Cape, Alexandria, Gibraltar, Dover."[20] However on July 25, 1909 Louis Blériot successfully made the first Channel crossing from Calais to Dover in an aeroplane. Blériot's crossing signalled the end of the Channel as a barrier-moat for England against foreign enemies.

Because the Kaiserliche Marine surface fleet could not match the British Grand Fleet, the Germans developed submarine warfare, which was to become a far greater threat to Britain. The Dover Patrol was set up just before war started to escort cross-Channel troopships and to prevent submarines from accessing the Channel, thereby obliging them to travel to the Atlantic via the much longer route around Scotland.

On land, the German army attempted to capture Channel ports (see "Race to the Sea"), but although the trenches are often said to have stretched "from the frontier of Switzerland to the English Channel", they reached the coast at the North Sea. Much of the British war effort in Flanders was a bloody but successful strategy to prevent the Germans reaching the Channel coast.

On 31 January 1917, the Germans restarted unrestricted submarine warfare leading to dire Admiralty predictions that submarines would defeat Britain by November,[21] the most dangerous situation Britain faced in either World War.

The Battle of Passchendaele in 1917 was fought to reduce the threat by capturing the submarine bases on the Belgian coast, though it was the introduction of convoys and not capture of the bases that averted defeat. In April 1918 the Dover Patrol carried out the famous Zeebrugge Raid against the U-boat bases. The naval blockade effected via the Channel and North Sea was one of the decisive factors in the German defeat in 1918.[22]

Second World War

During the Second World War, naval activity in the European theatre was primarily limited to the Atlantic. The early stages of the Battle of Britain[23] featured air attacks on Channel shipping and ports, and until the Normandy landings (with the exception of the Channel Dash) the narrow waters were too dangerous for major warships. Despite these early successes against shipping, the Germans did not win the air supremacy necessary for a cross Channel invasion.

The Channel subsequently became the stage for an intensive coastal war, featuring submarines, minesweepers, and Fast Attack Craft.[24]

British radar facilities during the Battle of Britain 1940

150 mm World War II German gun emplacement in Normandy.

Dieppe was the site of an ill-fated raid by Canadian and British armed forces. More successful was the later Operation Overlord (D-Day), a massive invasion of German-occupied France by Allied troops. Caen, Cherbourg, Carentan, Falaise and other Norman towns endured many casualties in the fight for the province, which continued until the closing of the so-called Falaise gap between Chambois and Montormel, then liberation of Le Havre.

As part of the Atlantic Wall, between 1940 and 1945 the occupying German forces and the Organisation Todt constructed fortifications round the coasts of the Channel Islands such as this observation tower at Les Landes, Jersey

The Channel Islands were the only part of the British Commonwealth occupied by Germany (excepting the part of Egypt occupied by the Afrika Korps at the time of the Second Battle of El Alamein, which was a protectorate and not part of the Commonwealth). The German occupation of 1940–1945 was harsh, with some island residents being taken for slave labour on the Continent; native Jews sent to concentration camps; partisan resistance and retribution; accusations of collaboration; and slave labour (primarily Russians and eastern Europeans) being brought to the islands to build fortifications. The Royal Navy blockaded the islands from time to time, particularly following the liberation of mainland Normandy in 1944. Intense negotiations resulted in some Red Cross humanitarian aid, but there was considerable hunger and privation during the occupation, particularly in the final months, when the population was close to starvation. The German troops on the islands surrendered on 9 May 1945, a few days after the final surrender in mainland Europe.

Population

The Channel is densely populated on both shores, with major ports and resorts having a combined population of over 3.5 million people. The most significant towns and cities along the Channel (each with more than 20,000 inhabitants, ranked in descending order; populations are the urban area populations from the 1999 French census, 2001 UK census, and 2001 Jersey census) are as follows:

England

- Brighton–Worthing–Littlehampton: 461,181 inhabitants, made up of:
  - Brighton: 155,919
  - Worthing: 96,964
  - Hove: 72,335
  - Littlehampton: 55,716
  - Lancing–Sompting: 30,360
- Portsmouth: 442,252, including
  - Gosport: 79,200
- Bournemouth & Poole: 383,713
- Southampton: 304,400
- Plymouth: 243,795
- Torbay (Torquay): 129,702
- Hastings–Bexhill: 126,386
- Eastbourne: 106,562
- Bognor Regis: 62,141
- Folkestone–Hythe: 60,039
- Weymouth: 56,043
- Dover: 39,078
- Exmouth: 32,972
- Falmouth–Penryn: 28,801

The walled city of Saint-Malo was a former stronghold of corsairs

- Ryde: 22,806
- St Austell: 22,658
- Seaford: 21,851
- Falmouth: 21,635
- Penzance: 20,255

France

- Le Havre: 248,547 inhabitants
- Calais: 104,852
- Boulogne-sur-Mer: 92,704
- Cherbourg: 89,704
- Saint-Brieuc: 85,849
- Saint-Malo: 50,675
- Lannion–Perros-Guirec: 48,990
- Dieppe: 42,202
- Morlaix: 35,996
- Dinard: 25,006
- Étaples–Le Touquet-Paris-Plage: 23,994
- Fécamp: 22,717

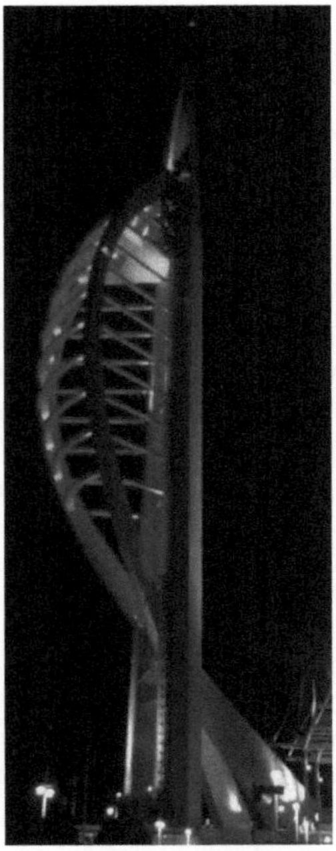

The Spinnaker Tower, Portsmouth at night, showing uplighting.

- Eu–Le Tréport: 22,019
- Trouville-sur-Mer–Deauville: 20,406
- Berck: 20,113

Channel Islands

- Saint Helier, Jersey: 28,310 inhabitants
- Saint Peter Port, Guernsey: 16,488 inhabitants
- Saint Anne, Alderney: 2,200 inhabitants
- Sark: 600 inhabitants
- Herm: 60 inhabitants

Shipping

The Channel has traffic on both the UK-Europe and North Sea-Atlantic routes, and is the world's busiest seaway, with over 500 ships per day.[25] Following an accident in January 1971 and a series of disastrous collisions with wreckage in February,[26] the Dover Traffic Separation System (TSS)[27] the world's first radar controlled TSS was set up by the International Maritime Organization. The scheme mandates that vessels travelling north must use the French side, travelling south the English side. There is a separation zone between the two lanes.[28]

In December 2002 the MV *Tricolor*, carrying £30m of luxury cars sank 32 km (20 mi) northwest of Dunkirk after collision in fog with the container ship *Kariba*. The cargo ship *Nicola* ran into the wreckage the next day. There was no loss of life.

The shore-based long range traffic control system was updated in 2003 and there is a series of Traffic Separation Systems in operation.[29] Though the system is inherently incapable of reaching the levels of safety obtained from aviation systems such as the Traffic Collision Avoidance System, it has reduced accidents to one or two per year.

Marine GPS systems allow ships to be preprogrammed to follow navigational channels accurately and automatically, further avoiding risk of running aground, but following the fatal collision between Dutch Aquamarine and Ash in October 2001, Britain's Marine Accident Investigation Branch (MAIB) issued a safety bulletin saying it believed that in these most unusual circumstances GPS use had actually contributed to the collision.[30] The ships were maintaining a very precise automated course, one directly behind the other, rather than making use of the full width of the traffic lanes as a human navigator would.

A combination of radar difficulties in monitoring areas near cliffs, a failure of a CCTV system, incorrect operation of the anchor, the inability of the crew to follow standard procedures of using a GPS to provide early warning of the ship dragging the anchor and reluctance to admit the mistake and start the engine led to the MV *Willy* running aground in Cawsand bay, Cornwall in January 2002. The MAIB report makes it clear that the harbour controllers were informed of impending disaster by shore observers before the crew were themselves aware.[31] The village of Kingsand was evacuated for three days because of the risk of explosion, and the ship was stranded for 11 days.[32] [33] [34]

Ecology

As a busy shipping lane, the Channel experiences environmental problems following accidents involving ships with toxic cargo and oil spills.[35] Indeed over 40% of the UK incidents threatening pollution occur in or very near the Channel.[36] One of the most infamous was the MSC *Napoli*, which with nearly 1700 tonnes of dangerous cargo was controversially beached in Lyme Bay, a protected World Heritage Site coastline. The ship had been damaged and was en route to Portland Harbour when much nearer harbours were available.

Transportation

Ferry

Important ferry routes are:

- Dover-Calais

The beach of Le Havre and a part of the rebuilt city

- Dover-Boulogne
- Dover-Dunkerque
- Newhaven-Dieppe
- Portsmouth-Caen (Ouistreham)
- Portsmouth-Cherbourg
- Portsmouth-Le Havre
- Portsmouth-Saint Malo
- Portsmouth-Jersey & Guernsey
- Poole-Saint Malo
- Poole-Cherbourg
- Weymouth-Saint Malo
- Plymouth-Roscoff

Channel Tunnel

Many travellers cross beneath the Channel using the Channel Tunnel, first proposed in the early 19th century and finally realised in 1994, connecting the UK and France by rail. It is now routine to travel between Paris or Brussels and London on the Eurostar train. Cars can also be carried on special trains between Folkestone and Calais.

Economy

Tourism

The coastal resorts of the Channel, such as Brighton and Deauville, inaugurated an era of aristocratic tourism in the early 19th century, which developed into the seaside tourism that has shaped resorts around the world. Short trips across the Channel for leisure purposes are often referred to as Channel Hopping.

Culture and languages

The two dominant cultures are English on the north shore of the Channel, French on the south. However, there are also a number of minority languages that are or were found on the shores and islands of the English Channel, which are listed here, with the Channel's name following them.

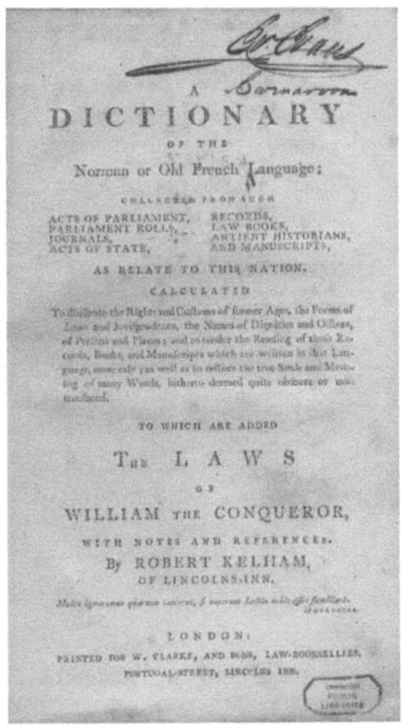
A DICTIONARY OF THE Norman or Old French Language; COLLECTED FROM SUCH ACTS OF PARLIAMENT, PARLIAMENT ROLLS, JOURNALS, ACTS OF STATE, RECORDS, LAW BOOKS, ANTIENT HISTORIANS, AND MANUSCRIPTS, AS RELATE TO THIS NATION. CALCULATED To illustrate the Rights and Customs of former Ages, the Forms of Law and Jurisprudence, the Names of Dignities and Offices, of Persons and Places; and to render the Reading of those Records, Books, and Manuscripts which are written in that Language, more easy; as well as to restore the true Sense and Meaning of many Words, hitherto deemed quite obscure or mistranslated.

TO WHICH ARE ADDED THE LAWS OF WILLIAM THE CONQUEROR, WITH NOTES AND REFERENCES. By ROBERT KELHAM, OF LINCOLNS-INN.

LONDON: PRINTED FOR W. CLARKE, AND SONS, LAW-BOOKSELLERS, PORTUGAL-STREET, LINCOLNS INN.

Kelham's *Dictionary of the Norman or Old French Language* (1779), dealing with England's Law French, a cross-Channel relic

Celtic Languages

- Breton - "Mor Breizh" (Sea of Brittany)
- Cornish - "Mor Bretannek"

Germanic languages

- English
- Dutch - "het Kanaal" (the Channel)
- German - "der Ärmelkanal" (the Sleeve Channel)

Dutch previously had a larger range, and extended into parts of modern-day France. For more information, please see French Flemish.

Romance languages

- French - "La Manche"
- Gallo
- Norman, including the Channel Island vernaculars:
  - Anglo-Norman (extinct, but fossilised in certain English law phrases)
  - Auregnais (extinct)
  - Cotentinais - "Maunche"
  - Guernesiais - "Ch'nal"
  - Jèrriais - "Ch'na"
  - Sercquais
- Picard

Most other languages tend towards variants of the French and English forms, but notably Welsh has "Môr Udd".

Notable channel crossings

As one of the narrowest but most famous international waterways lacking dangerous currents, the Channel has been the first objective of numerous innovative sea, air, and human powered crossing technologies.

Date	Crossing	Participant(s)	Notes
7 January 1785	First crossing by air (in balloon, from Dover to Calais)	Jean Pierre François Blanchard (France) John Jeffries (US)	—
15 June 1785	First air crash (in combination hydrogen / hot-air balloon)	Pilâtre de Rozier (France) Pierre Romain (France)	Attempted crossing similar to Blanchard/Jeffries
March 1816	the French paddle steamer *Elise* (ex Scottish-built Margery or Margory) was the first steamer to cross the Channel.		
9 May 1816	Paddle steamer *Defiance*, Captain William Wager, was the first steamer to cross the Channel to Holland [37]		
10 June 1821	Paddle steamer *Rob Roy*, first passenger ferry to cross channel.		The steamer was purchased subsequently by the French postal administration and renamed *Henri IV*.
June 1843	First ferry connection through Folkestone-Boulogne		Commanding officer Captain Hayward
25 August 1875	First person known to swim the channel (Dover to Calais, 21 hrs, 45 min)	Matthew Webb (UK)	Attempted crossing on 12 August the same year; forced to abandon swim because of strong winds/rough sea conditions
27 March 1899	First radio transmission across the Channel (from Wimereux to South Foreland Lighthouse)	Guglielmo Marconi (Italy)	
25 July 1909	First person to cross the channel in a heavier-than-air aircraft (the *Blériot XI*) (Calais to Dover, 37 minutes)	Louis Blériot (France)	Encouraged by £1000 prize offered by the *Daily Mail* for first successful flight across the Channel
2 June 1910	First person to make a double crossing of the Channel in a heavier-than-air aircraft (a Short Wright biplane).[38]	Charles Stewart Rolls (UK)	From Swingfield Downs, Kent to Sangatte, France, returning to Eastchurch, Kent.
23 August 1910	First aircraft flight with passengers	John Bevins Moisant (US)	Passengers were mechanic Albert Fileux and Moisant's cat.
16 April 1912	First woman to fly across the Channel (Dover to Calais, 59 minutes)	Harriet Quimby (US)	Her accomplishment did not receive much media attention, as the RMS Titanic sank the evening before.
23 August 1926	First woman to swim across the channel (Cap Gris Nez to Kingsdown, 14 hours 39 minutes)	Gertrude Ederle (US)	Five men had swum the channel before Ederle. Ederle beat their best time by two hours, creating a record for a female swimmer that stood until Florence Chadwick swam it in 13 hours 20 minutes in 1950.[39]
18 September 1928	First flight across the Channel by autogyro	Juan de la Cierva (SPA)	Achieved as part of the first flight by autogyro between London and Paris.[40]

19 June 1931	First crossing in a glider	Lissant Beardmore (UK)	Aero-tow from Lympne to an altitude of 14000 feet (**unknown operator: u'strong'** m). Landed at Saint-Inglevert Airfield, Pas-de-Calais.[41]
25 July 1959	Hovercraft crossing (Calais to Dover, 2 hours 3 minutes)	SR-N1	Sir Christopher Cockerell was on board
22 August 1972	First solo hovercraft crossing (same route as SR-N1; 2 hours 20 minutes[42])	Nigel Beale (UK)	
1974	Coracle (13½ hours)	Bernard Thomas (UK)	As part of a publicity stunt, the journey was undertaken to demonstrate how the Bull Boats of the Mandan Indians of North Dakota could have been copied from Welsh coracles introduced by Prince Madog in the 12th century.[43]
12 June 1979	First human-powered aircraft to fly over the Channel (in 55-pound (25 kg) *Gossamer Albatross*)	Bryan Allen (US)	Won a £100,000 Kremer Prize; Allen pedalled for three hours.
14 September 1995	Fastest crossing by hovercraft, 22 minutes by "Princess Anne"	MCH SR-N4 MkIII	Craft was designed as a ferry
1997	First vessel to complete a solar-powered crossing using photovoltaic cells.	SB *Collinda*	—
14 June 2004	New record time for crossing in amphibious vehicle (the Gibbs Aquada, two-seater open-top sports car)	Richard Branson (UK)	Completed crossing in 1 hour 40 minutes 6 seconds - previous record was 6 hours.
31 July 2003	Crossing in a 20-mile (**unknown operator: u'strong'** km) long freefall using a wingsuit and a carbon fibre wing	Felix Baumgartner (Austria)	
26 July 2006	New record time for crossing in hydrofoil car (the Rinspeed Splash, two-seater open-top sports car)	Frank M. Rinderknecht (SUI)	Completed crossing in 3 hours 14 minutes[44]
25 September 2006	First crossing on a towed inflatable object (not a powered inflatable boat)	Stephen Preston (UK)	Completed crossing in 180 min[45]
July 2007	BBC Top Gear presenters *drive* to France in amphibious cars.	Jeremy Clarkson, Richard Hammond, James May (UK)	Completed the crossing in a 1996 Nissan D21 pick-up (the "Nissank"), fitted with a Honda outboard engine.[46]
26 September 2008	First crossing with a jetpack	Yves Rossy (SUI)	Crossing completed in less than ten minutes.[47]
12 March 2010	First crossing by water ski	Christine Bleakley (UK)	Completed in just over 1 hour 40 minutes. She completed the challenge for BBC Sport Relief, falling eight times during the crossing.
28 May 2010	First crossing by helium balloon	Jonathan Trappe (US)	Completed in 4 hours. He crossed the Channel dangling beneath a cloud of coloured helium balloons and controlled his altitude by cutting the balloons free one by one, with a pair of scissors.[48]

18 July 2010	First crossing by RS Tera	Children aged 7 to 16 from CCSC, DSC and Downs SC (UK)	Crossing made in single-handed 9-foot (**unknown operator: u'strong'** m) sailing dinghies. Dover to Boulogne (27 miles (**unknown operator: u'strong'** km)) in 5 hours.
20 August 2011	First Crossing by Sea Scooters	A four-man relay team from Scarborough, North Yorkshire, headed by Heath Samples LL.B Hons, crossed from Shakespeare Beach to Wissant.	It took 12 hours 26 minutes 39 seconds. It set a new Guinness World Record.

By boat

Pierre Andriel crossed the English Channel aboard the *Élise*, ex the Scottish p.s. "Margery" in March 1816, one of the earliest sea going voyages by steam ship.

The paddle steamer *Defiance*, Captain William Wager, was the first steamer to cross the Channel to Holland, arriving there on 9 May 1816.[37]

On June 10, 1821, English-built paddle steamer *Rob Roy* was the first passenger ferry to cross channel. The steamer was purchased subsequently by the French postal administration and renamed *Henri IV* and put into regular passenger service a year later. It was able to make the journey across the Straits of Dover in around three hours.[49]

In June 1843, because of difficulties with Dover harbour, the South Eastern Railway company developed the Boulogne-sur-Mer-Folkestone route as an alternative to Calais-Dover. The first ferry crossed under the command of Captain Hayward.[50]

The Mountbatten class hovercraft (MCH) entered commercial service in August 1968, initially between Dover and Boulogne but later also Ramsgate (Pegwell Bay) to Calais. The journey time Dover to Boulogne was roughly 35 minutes, with six trips per day at peak times. The fastest crossing of the English Channel by a commercial car-carrying hovercraft was 22 minutes, recorded by the *Princess Anne* MCH SR-N4 Mk3 on 14 September 1995,[51] for the 10:00 am service .

The youngest recorded sailors to cross the Channel were a team of eight with ages ranging from 7 to 16 on 18 July 2010. They sailed their single-handed RS Tera dinghies 27 miles (**unknown operator: u'strong'** km) from Dover to Boulogne in 5 hours and were tracked by Dover Coastguard Radar, who retain a record of the passage. They were from three UK clubs: Castle Cove SC, Dabchicks SC and Downs SC.

By swimming

The sport of Channel swimming traces its origins to the latter part of the 19th century when Captain Matthew Webb made the first observed and unassisted swim across the Strait of Dover, swimming from England to France on 24–25 August 1875 in 21 hours 45 minutes.

In 1927, at a time when fewer than ten swimmers had managed to emulate the feat and many dubious claims were being made, the Channel Swimming Association (CSA) was founded to authenticate and ratify swimmers' claims to have swum the Channel and to verify crossing times. The CSA was dissolved in 1999 and was succeeded by two separate organisations: CSA (Ltd) and the Channel Swimming and Piloting Federation (CSPF). Both observe and authenticate cross-Channel swims in the Strait of Dover. The Channel Crossing Association was set up at about this time to cater for unorthodox crossings.

The team with the most number of Channel swims to its credit is the International Sri Chinmoy Marathon Team, with 35 crossings by 25 members (by 2005).[52]

By the end of 2005, 811 people had completed 1,185 verified crossings under the rules of the CSA, the CSA (Ltd), the CSPF and Butlins.

The number of swims conducted under and ratified by the Channel Swimming Association to 2005 was 982 by 665 people. This includes 24 two-way crossings and three three-way crossings.

The number of ratified swims to 2004 was 948 by 675 people (456 men, 214 women). There have been 16 two-way crossings (9 by men and 7 by women). There have been three three-way crossings (2 by men and 1 by a woman). (It is unclear whether this last set of data is comprehensive or CSA only.)

The Strait of Dover is the busiest stretch of water in the world. It is governed by International Law as described in this document.[53] It states: "However, in exceptional cases the French Maritime Authorities may grant authority for unorthodox craft to cross French territorial waters within the Traffic Separation Scheme when these craft set off from the British coast, on condition that the request for authorisation is sent to them with the opinion of the British Maritime Authorities." It is therefore theoretically possible to hire a non CSA or CS&PF pilot boat when swimming the channel, although it would be difficult to convince the MCA to endorse the trip.

The CCA, CSA, and CS&PF are the organisations escorting channel swims, because their pilots have the experience, qualifications, and equipment to guarantee the safety of the swimmers they escort.

The fastest verified swim of the Channel was by the Bulgarian Petar Stoychev on 24 August 2007, in 6 hours 57 minutes 50 seconds.

By car

On 16 September 1965, two Amphicars crossed from Dover to Calais. One was crewed by two Army Officers, Captain Mike Bailey REME and Captain Peter Tappenden RAOC, the other by Tim Dill-Russell and Sgt Joe Minto RASC. The crossing took 7 hours 20 minutes. Conditions in mid channel were up to force 5. The cars went on to the Frankfurt Motor Show of that year where they were put on display.[54]

In 2007, the presenters of the BBC programme Top Gear (Jeremy Clarkson, Richard Hammond and James May) "drove" across the Channel from England to France. They did it by designing 'Amphibious Cars' which could be driven on land and also operate in water. After four attempts - twice failing to leave Dover Harbour - they reached the coast of France in a Nissan pick-up (The Nissank) with an outboard motor and oil drums attached to the back to aid stability in the open water.[46] The other two vehicles that attempted the crossing (a Triumph Herald with a sail and a Volkswagen Campervan with a propeller attached to the flywheel) both sank.

Clarkson believed it might be possible to break the world record for crossing the channel in this manner, but the team was unsuccessful.[55]

The Daily Mail claimed that the BBC received criticism from the coastguard who claimed that they had not been told that the stunt was going to take place, and allegedly branded it "completely irresponsible", despite the aired episode showing the co-operation of the coastguard.[56]

See also

- Doggerland
- Booze cruise
- Phoenix breakwaters
- List of firsts in aviation
- List of successful English Channel swimmers

References

[1] "English Channel". *The Columbia Encyclopedia*, 2004.

[2] "English Channel." Encyclopædia Britannica 2007.

[3] "Limits of Oceans and Seas, 3rd edition + corrections" (http://www.iho-ohi.net/iho_pubs/standard/S-23/S23_1953.pdf). International Hydrographic Organization. 1971. pp. 42 [corrections to page 13] and 6. . Retrieved 25 September 2010.

[4] "English Channel." *The Hutchinson Unabridged Encyclopedia including Atlas.* 2005.

[5] File:Allied Invasion Force.jpg + French map of Channel

[6] Gupta, Sanjeev; Jenny S. Collier, Andy Palmer-Felgate & Graeme Potter (2007). "Catastrophic flooding origin of shelf valley systems in the English Channel" (http://www.nature.com/nature/journal/v448/n7151/full/nature06018.html). *Nature* **448** (7151): 342–345. Bibcode 2007Natur.448..342G. doi:10.1038/nature06018. PMID 17637667. . Retrieved 2007-07-18. Lay summary (http://www.msnbc.msn.com/id/19833064/) – *msnbc.com* (2007-07-18).

[7] "GPG Cambridge.ac Physics Today, Sonar mapping suggests that the English Channel was created by two megafloods, (extract of Gupta Potter), Freely downloadabe PDF" (http://www.qpg.geog.cam.ac.uk/research/projects/englishchannelfloods/physicstoday.pdf) (PDF). . Retrieved 2010-04-27.

[8] Professor Bryony Coles. "The Doggerland project" (http://humanities.exeter.ac.uk/archaeology/research/projects/title,89282,en.html). University of Exeter. . Retrieved 3 January 2011.

[9] "Jonathan Potter: Map : The British Channel" (http://www.jpmaps.co.uk/map/id.22553). Jpmaps.co.uk. . Retrieved 2010-04-27.

[10] "A chart of the British Channel, Jefferys, Thomas, 1787" (http://www.davidrumsey.com/maps6489.html). Davidrumsey.com. 1999-02-22. . Retrieved 2010-04-27.

[11] " Map Of Great Britain (http://www.collectbritain.co.uk), Ca. 1450", Collect Britain

[12] Room A. *Placenames of the world: origins and meanings*, p. 6.

[13] cf. "Kernow", the Cornish for Cornwall.

[14] Hermann Flohn, Roberto Fantechi, *The Climate of Europe, past, present, and future*, 1984, ISBN 90-277-1745-1, p.46 (http://books.google.co.uk/books?id=w0pAEI6AWLgC&pg=PA46&lpg=PA46&dq=1684+channel+ice&source=bl&ots=Md4yEs4n4U&sig=FGTVc3FnwOv5wGz-q6YJduG7PS0&hl=en&ei=Gr_4TJSnOJK1hAfAufihCQ&sa=X&oi=book_result&ct=result&resnum=3&ved=0CCMQ6AEwAg#v=onepage&q=1684 channel ice&f=false)

[15] PastPresented,info: The Great Frost of 1683-4 (http://www.pastpresented.info/frost1683.htm)

[16] "History Compass" (http://www.history-compass.com/images/store/HICO/chapters/523.pdf). History Compass. . Retrieved 2010-04-27.

[17] "Germany The migration period" (http://www.britannica.com/eb/article-58084/Germany). . Retrieved 2007-07-24.

[18] Nick Attwood MA. "The Holy Island of Lindisfarne - The Viking Attack" (http://www.lindisfarne.org.uk/793/index.htm). . Retrieved 2007-07-24.

[19] britishbattles.com (2007). "The Spanish Armada: Sir Francis Drake" (http://www.britishbattles.com/spanish-war/spanish-armada.htm). . Retrieved 2007-07-24.

[20] Geoffrey Miller. *The Millstone: Chapter 2* (http://www.manorhouse.clara.net/book3/chapter2.htm). . Retrieved 2008-11-01. quoting Fisher, *Naval Necessities I*, p. 219

[21] "U-Boat warfare at the Atlantic during World War I" (http://web.archive.org/web/20080612084614/http://www.germannotes.com/hist_ww1_uboat.shtml). German Notes. Archived from the original (http://www.germannotes.com/hist_ww1_uboat.shtml) on 2008-06-12. . Retrieved 2008-11-01.

[22] "His Imperial German Majesty's U-boats in WWI: 6. Finale" (http://uboat.net/history/wwi/part6.htm). uboat.net. . Retrieved 2009-09-13.

[23] "Fact File: Battle of Britain" (http://www.bbc.co.uk/ww2peopleswar/timeline/factfiles/nonflash/a1057330.shtml?sectionId=2&articleId=1057330). BBC. . Retrieved 2008-11-01.

[24] Campaigns of World War II, Naval History Homepage. "Atlantic, WW2, U-boats, convoys, OA, OB, SL, HX, HG, Halifax, RCN ..." (http://www.naval-history.net/WW2CampaignsStartEurope.htm). . Retrieved 2007-07-24.

[25] "The Dover Strait" (http://www.mcga.gov.uk/c4mca/mcga07-home/emergencyresponse/mcga-searchandrescue/mcga-hmcgsar-sarsystem/channel_navigation_information_service__cnis_/the_dover_strait.htm). Maritime and Coastguard Agency. 2007. . Retrieved 2008-10-08.

[26] "History of CNIS" (http://www.mcga.gov.uk/c4mca/mcga07-home/emergencyresponse/mcga-searchandrescue/mcga-hmcgsar-sarsystem/channel_navigation_information_service__cnis_/history_of_cnis.htm). Maritime and Coastguard Agency. 2007. . Retrieved 2008-11-01.

[27] "Dover Strait TSS" (http://www.mcga.gov.uk/c4mca/mcga07-home/emergencyresponse/mcga-searchandrescue/mcga-hmcgsar-sarsystem/channel_navigation_information_service__cnis_/dops_-_dover_strait_tss_chartlet.htm). Maritime and Coastguard Agency. . Retrieved 2008-11-01.

[28] "World Marine Guide - English Channel" (http://www.worldmarineguide.com/marina/united-kingdom/uk-south-coast/article-english-channel). . Retrieved 5 August 2011.

[29] Chartlets published by the Maritime and Coastguard Agency (http://www.mcga.gov.uk/c4mca/mcga07-home/emergencyresponse/mcga-searchandrescue/mcga-hmcgsar-sarsystem/channel_navigation_information_service__cnis_/dops_-_dover_strait_tss_chartlet.htm)

[30] "Safety Bulletin 2" (http://www.maib.gov.uk/cms_resources/SB_ 2_2001_ Ash_and_Dutch_Aquamarin.pdf) (PDF). Marine Accident Investigation Branch. 2001. . Retrieved 2008-11-01.

[31] "Report on the Investigation of the grounding of MV *Willy*" (http://www.maib.gov.uk/cms_resources/willy.pdf) (PDF). Marine Accident Investigation Branch. October 2002. . Retrieved 2008-11-01.

[32] "Picture gallery: Cornwall's stranded tanker" (http://news.bbc.co.uk/1/hi/england/1742910.stm). London: BBC. 5 January 2002. . Retrieved 2008-11-01.

[33] "Salvage team hunts for leak" (http://news.bbc.co.uk/1/hi/england/1745945.stm). London: BBC. 6 January 2002. . Retrieved 2008-11-01.

[34] "Stranded tanker safe in port" (http://news.bbc.co.uk/1/hi/england/1759670.stm). London: BBC. 14 January 2002. . Retrieved 2008-11-01.

[35] "Tanker wreck starts leaking oil" (http://news.bbc.co.uk/1/hi/world/europe/guernsey/4668664.stm). London: BBC. 1 February 2006. . Retrieved 2008-11-01.

[36] "Annual Survey of Reported Discharges" (http://www.mcga.gov.uk/c4mca/pacops_final_report_2006.pdf). Maritime and Coastguard Agency. 2006. . Retrieved 2008-11-01.

[37] Dawson, Charles (February 1998). "P. S. Defiance, the first steamer to Holland, 9 May 1816". *The Mariner's Mirror* (The Society for Nautical Research) **84** (1): 84.

[38] Humphreys, Roy (2001). *Kent Aviation, a Century of Flight*. Stroud: Sutton Publishing. p. 189. ISBN 0-7509-2790-9.

[39] Severo, Richard (December 1, 2003). "Gertrude Ederle, the First Woman to Swim Across the English Channel, Dies at 98" (http://www.nytimes.com/2003/12/01/sports/gertrude-ederle-the-first-woman-to-swim-across-the-english-channel-dies-at-98.html). *New York Times*. . Retrieved 2009-08-11. "Gertrude Ederle, who was called *America's best girl* by President Calvin Coolidge in 1926 after she became the first woman to swim across the English Channel, died yesterday at a nursing home in Wyckoff, N.J. She was 98."

[40] "Channel Flight By Autogiro. Spanish Airman's Success." *The Times* (London). Wednesday, 19 September 1928. Issue **45002**, col F, p. 14.

[41] "Channel Crossed by Glider." *The Times* (London). Saturday, 20 June 1931. Issue **45854**, col F, p. 12.

[42] Verifiable in Hovercraft Club of Great Britain Records and Archives.

[43] "Wales on Britannia: Facts About Wales & the Welsh" (http://www.britannia.com/celtic/wales/facts/facts7.html). Britannia.com. . Retrieved 2010-04-27.

[44] Stuart Waterman (July 27, 2006). "Rinspeed "Splash" sets English Channel record" (http://www.autoblog.com/2006/07/27/rinspeed-splash-sets-english-channel-record/). Autoblog. . Retrieved 2008-11-01.

[45] "Inflatable Drag" (http://www.stupidsteve.co.uk/inflatable.html). . Retrieved 2008-11-01.

[46] "1996 Nissan Truck [D21] in "Top Gear, 2002–2010"" (http://imcdb.org/vehicle_132991-Nissan-Pickup-D21-1996.html). IMCDb.org. . Retrieved 2010-04-27.

[47] "Pilot completes jetpack challenge" (http://news.bbc.co.uk/1/hi/uk/7637327.stm). London: BBC. 26 September 2008. . Retrieved 2008-11-01.

[48] American crosses Channel carried by helium balloons (http://www.guardian.co.uk/uk/2010/may/28/american-channel-helium-balloons), *The Guardian*, accessed 28 May 2010

[49] The History of the Channel Ferry (http://www.sailingandboating.co.uk/history-channel-ferry.html)

[50] (http://www.theotherside.co.uk/tm-heritage/background/ferries.htm) Channel ferries & ferry ports

[51] "Hovercraft deal opens show" (http://news.bbc.co.uk/onthisday/hi/dates/stories/june/15/newsid_3025000/3025267.stm). London: BBC News. 15 June 1966. . Retrieved 2008-11-01.

[52] "Sri Chinmoy Marathon Team Channel Swim List" (http://www.srichinmoyraces.org/channel/channel_swimmers/channel_swimmers_list). . Retrieved 2008-11-01.

[53] "Unorthodox Crossing of the Dover Strait Traffic Separation Scheme" (http://www.mcga.gov.uk/c4mca/mcga07-home/emergencyresponse/mcga-searchandrescue/mcga-hmcgsar-sarsystem/channel_navigation_information_service__cnis_/dops_-_all-sar_cnis_unorthodox_crossings.htm). Maritime and Coastguard Agency. . Retrieved 2008-11-01.

[54] Autocar article entitled Cars Ahoy published 10 December 1965

[55] BBC Top Gear Series 10 Episode 2

[56] "Coastguards' fury as Top Gear stars attempt to 'drive' across the Channel | Mail Online" (http://www.dailymail.co.uk/tvshowbiz/article-469782/Coastguards-fury-Top-Gear-stars-attempt-drive-Channel.html). Dailymail.co.uk. 2007-07-20. . Retrieved 2010-04-27.

External links

- Full Channel swim lists and swimmer information (http://www.dover.uk.com/channelswimming/)
- Oceanus Britannicus or British Sea (http://freepages.genealogy.rootsweb.com/~genmaps/genfiles/COU_files/ENG/aaEng/ortelius_anglia-epitome-det3_1595.jpg)
- Channel swimmers website (http://www.thechannelswimmers.com)
- Archives of long distance swimming (http://www.soloswims.com)
- Sponsor David Walliam's Sport Relief swim (http://www.sportrelief.com/superswim/)
- Channel Swimming and Piloting Federation (http://www.channelswimming.net/)
- Channel Swimming Association (http://www.channelswimmingassociation.com/)
- World War II Eye Witness Account - Audio Recording (http://www.readagainbooks.com/airbattleovertheenglishchannel.htm) Air Battle over the English Channel (1940)

Mediterranean_Sea

Mediterranean Sea		
[[Image:Mediterranean Sea political map-en.svg		**Mediterranean Sea**} - Composite satellite image of the Mediterranean Sea]] Composite satellite image of the Mediterranean Sea
Coordinates	35°N 18°E	
Basin countries	Albania, Algeria, Andorra, Austria, Belarus, Bosnia and Herzegovina, Bulgaria, Burundi, Central African Republic, Chad, (Republic of the) Congo, Croatia, Cyprus, Czech Republic, Egypt, Eritrea, Ethiopia, France, Georgia, Germany, Gibraltar, Greece, Hungary, Israel, Italy, Kenya, Kosovo (independence disputed, claimed by Serbia), Lebanon, Libya, Liechtenstein, (Republic of) Macedonia, Malta, Moldova, Monaco, Montenegro, Morocco, Niger, Palestinian territories, Poland,[1] Romania, Russia, Rwanda, San Marino, Slovakia, Slovenia, South Sudan, Spain, Sudan, Switzerland, Syria, Tanzania, Tunisia, Turkey, Uganda, Ukraine, Vatican City	
Surface area	2500000 km^2 (**unknown operator: u'strong'** sq mi)	
Average depth	1500 m (**unknown operator: u'strong'** ft)	
Max depth	5267 m (**unknown operator: u'strong'** ft)	
Residence time (of Ocean water)	80-100 years[2]	
Islands	3300+	

The **Mediterranean Sea** is a sea connected to the Atlantic Ocean surrounded by the Mediterranean region and almost completely enclosed by land: on the north by Europe and Anatolia, on the south by North Africa, and on the east by the Levant. The sea is sometimes considered a part of the Atlantic Ocean, although it is often identified as a completely separate body of water.

The name Mediterranean is derived from the Latin *mediterraneus*, meaning "inland" or "in the middle of the land" (from *medius*, "middle" and *terra*, "land"). It covers an approximate area of 2.5 million km^2 (965,000 sq mi), but its connection to the Atlantic (the Strait of Gibraltar) is only 14 km (**unknown operator: u'strong'** mi) wide. In oceanography, it is sometimes called the *Eurafrican Mediterranean Sea* or the *European Mediterranean Sea* to distinguish it from mediterranean seas elsewhere.[3] [4]

The Mediterranean Sea has an average depth of 1500 m (**unknown operator: u'strong'** ft) and the deepest recorded point is 5267 m (**unknown operator: u'strong'** ft) in the Calypso Deep in the Ionian Sea.

It was an important route for merchants and travelers of ancient times that allowed for trade and cultural exchange between emergent peoples of the region — the Mesopotamian, Persian, Egyptian, Phoenician, Carthaginian, Iberian, Greek and Byzantine, Roman and Italian, Illyrian and Albanian, Thracian, Levantine, Gallic, Arabic, Berber, Jewish, Aragonese, Nubian, Slavic and Turkish cultures. The history of the Mediterranean region is crucial to understanding the origins and development of many modern societies. "For the three quarters of the globe, the Mediterranean Sea is similarly the uniting element and the centre of World History."[5]

Name

The term *Mediterranean* derives from the Latin word *mediterraneus*, meaning "in the middle of earth" or "between lands" (*medius*, "middle, between" + *terra*, "land, earth"): as it is between the continents of Africa and Europe. The Greek name *Mesogeios* (Μεσόγειος), is similarly from μέσο, "middle" + γη, "land, earth").[6]

The Mediterranean Sea has historically had several names. For example the Romans commonly called it *Mare Nostrum* (Latin, "Our Sea"), and occasionally *Mare Internum* (Sallust, *Jug.* 17).

In the Bible, it was primarily known as the "Great Sea" (Num. 34:6,7; Josh. 1:4, 9:1, 15:47; Ezek. 47:10,15,20), or simply "The Sea" (1 Kings 5:9; *comp.* 1 Macc. 14:34, 15:11); however, it has also been called the "Hinder Sea", due to its location on the west coast of the Holy Land, and therefore behind a person facing the east, as referenced in the Old Testament, sometimes translated as "Western Sea", (Deut. 11:24; Joel 2:20). Another name was the "Sea of the Philistines" (Exod. 23:31), from the people occupying a large portion of its shores near the Israelites.

In Modern Hebrew, it has been called *HaYyam HaTtikhon* (הַיָּם הַתִּיכוֹן), "the middle sea", a literal adaptation of the German equivalent *Mittelmeer*. In Turkish, it is known as *Akdeniz*, "the white sea". In modern Arabic, it is known as *al-Baḥr al-Abyaḍ al-Mutawassiṭ* (البحر الأبيض المتوسط), "the White Middle Sea", while in Islamic and older Arabic literature, it was referenced as *Baḥr al-Rūm* (بحر الروم), or "the Roman/Byzantine Sea."

History

Several ancient civilisations were located around its shores; thus it has had a major influence on those cultures. It provided routes for trade, colonisation and war, and provided food (by fishing and the gathering of other seafood) for numerous communities throughout the ages.[7]

The Holy League naval victory over the Ottoman Empire at the *Battle of Lepanto*, 1571, the last battle to be fought primarily between galleys.

The sharing of similar climate, geology and access to a common sea led to numerous historical and cultural connections between the ancient and modern societies around the Mediterranean.

Two of the most notable Mediterranean civilisations in classical antiquity were the Greek city states and the Phoenicians. When Augustus founded the Roman Empire, the Mediterranean Sea began to be called *Mare Nostrum* (literally:"Our Sea") by the Romans.

Darius I of Persia, who conquered Ancient Egypt, built a canal linking the Mediterranean to the Red Sea. Darius's canal was wide enough for two triremes to pass each other with oars extended, and required four days to traverse.[8]

The western Roman empire collapsed around AD 476. Temporarily the east was again dominant as the Byzantine Empire formed from the eastern half of the Roman empire. Another power soon arose in the east: Islam. At its greatest extent, the Arab Empire controlled 75% of the Mediterranean region.

Europe was reviving, however, as more organized and centralized states began to form in the later Middle Ages after the Renaissance of the 12th century.

Ottoman power continued to grow, and in 1453, the Byzantine Empire was extinguished with the fall of Constantinople. The growing naval prowess of the European powers confronted further rapid Ottoman expansion in the region when the Battle of Lepanto checked the power of the Ottoman navy. The development of oceanic shipping began to affect the entire Mediterranean. Once, all trade from the east had passed through the region, but now the circumnavigation of Africa allowed spices and other goods to be imported through the Atlantic ports of western Europe.[9] [10] [11]

Geography

A satellite image showing the Strait of Gibraltar. Left is the Iberian Peninsula in Europe; on the right, the Maghreb in Africa.

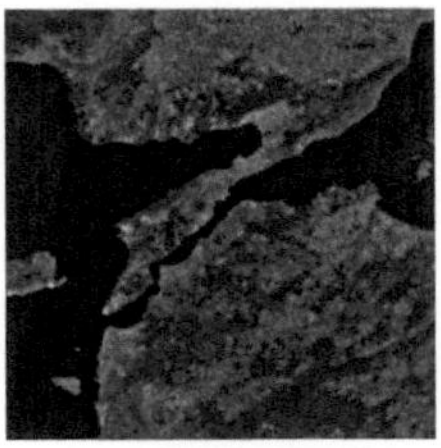

The Dardanelles strait. The north side is Europe with the Gelibolu Peninsula in the Thrace region; the south side is Anatolia in Asia.

The Mediterranean Sea is connected to the Atlantic Ocean by the Strait of Gibraltar in the west and to the Sea of Marmara and the Black Sea, by the Dardanelles and the Bosporus respectively, in the east. The Sea of Marmara is often considered a part of the Mediterranean Sea, whereas the Black Sea is generally not. The 163 km (**unknown operator: u'strong'** mi) long man-made Suez Canal in the southeast connects the Mediterranean Sea to the Red Sea.

Large islands in the Mediterranean include Cyprus, Crete, Euboea, Rhodes, Lesbos, Chios, Kefalonia, Corfu, Naxos and Andros in the eastern Mediterranean; Sardinia, Corsica, Sicily, Cres, Krk, Brač, Hvar, Pag, Korčula and Malta in the central Mediterranean; and Ibiza, Majorca and Minorca (the Balearic Islands) in the western Mediterranean.

The typical Mediterranean climate has hot, dry summers and mild, rainy winters. Crops of the region include olives, grapes, oranges, tangerines, and cork.

Extent

The International Hydrographic Organization defines the limits of the Mediterranean Sea as follows:[12]

Stretching from the Strait of Gibraltar in the West to the entrances to the Dardanelles and the Suez Canal in the East, the Mediterranean Sea is bounded by the coasts of Europe, Africa and Asia, and is divided into two deep basins:

- Western Basin:
  - On the west: A line joining the extremities of Cape Trafalgar (Spain) and Cape Spartel (Africa).
  - On the northeast: The West Coast of Italy. In the Strait of Messina a line joining the North extreme of Cape Paci (15°42'E) with Cape Peloro, the East extreme of the Island of Sicily. The North Coast of Sicily.
  - On the east: A line joining Cape Lilibeo the Western point of Sicily (37°47′N 12°22′E), through the Adventure Bank to Cape Bon (Tunisia).
- Eastern Basin:
  - On the west: The Northeastern and Eastern limits of the Western Basin.
  - On the northeast: A line joining Kum Kale (26°11'E) and Cape Helles, the Western entrance to the Dardanelles.
  - On the southeast: The entrance to the Suez Canal.

- On the east: The coasts of Syria and Palestine.

(It should be noted that the coast referred to as belonging to Palestine in this document dating to 1953 has been within the internationally recognised borders of the country known as Israel since 1948. Of the territories administered by the Palestinian Authority, only the Gaza Strip has a sea coast.)

Oceanography

Being nearly landlocked affects conditions in the Mediterranean Sea: for instance, tides are very limited as a result of the narrow connection with the Atlantic Ocean. The Mediterranean is characterized and immediately recognised by its deep blue colour.

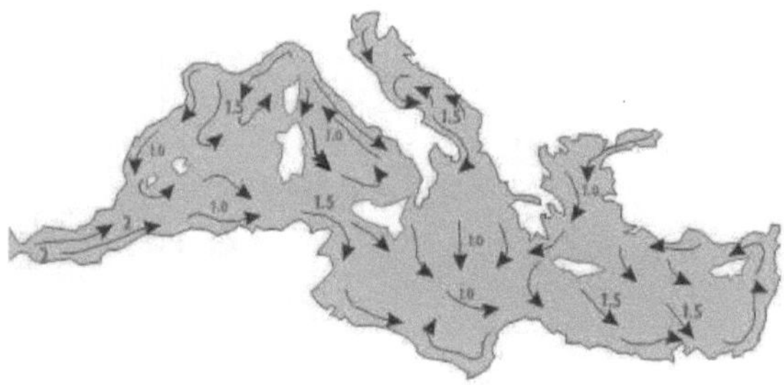

Predominant surface currents for June

Evaporation greatly exceeds precipitation and river runoff in the Mediterranean, a fact that is central to the water circulation within the basin.[13] Evaporation is especially high in its eastern half, causing the water level to decrease and salinity to increase eastward.[14] This pressure gradient pushes relatively cool, low-salinity water from the Atlantic across the basin; it warms and becomes saltier as it travels east, then sinks in the region of the Levant and circulates westward, to spill over the Strait of Gibraltar.[15] Thus, seawater flow is eastward in the Strait's surface waters, and westward below; once in the Atlantic, this chemically distinct Mediterranean Intermediate Water can persist thousands of kilometres away from its source.[16]

Coastal countries

Twenty-one modern states have a coastline on the Mediterranean Sea. They are:

Map of the Mediterranean Sea

- **Europe** (from west to east): Spain, France, Monaco, Italy, Malta, Slovenia, Croatia, Bosnia and Herzegovina, Montenegro, Albania, Greece and Turkey (East Thrace)
- **Asia** (from north to south): Turkey (Anatolia), Cyprus, Syria, Lebanon, Israel, Egypt (the Sinai Peninsula)
- **Africa** (from east to west): Egypt, Libya, Tunisia, Algeria and Morocco

Turkey and Egypt are transcontinental countries. The southernmost islands of Italy, the Pelagie islands, are geologically part of the African continent.

Several other territories also border the Mediterranean Sea (from west to east):

- British overseas territory of Gibraltar

- Spanish enclaves of Ceuta and Melilla and nearby islands
- British sovereign base area of Akrotiri and Dhekelia
- Turkish Republic of Northern Cyprus (disputed)
- Gaza Strip of the Palestinian Territories

Capital cities of sovereign countries and major cities (municipalities) with populations larger than 200,000 people bordering the Mediterranean Sea are (capitals with fewer than 200,000 inhabitants are marked with an asterisk):

Country	Cities
Spain	Alicante, Badalona, Barcelona, Cartagena, Málaga, Palma, Valencia
France	Marseille, Montpellier, Nice
Monaco	Monaco*
Italy	Bari, Catania, Genoa, Messina, Naples, Palermo, Rome, Trieste, Venice
Malta	Valletta*
Albania	Durrës
Greece	Athens, Patras, Thessaloniki
Cyprus	Limassol
Turkey	Antalya, Iskenderun, Izmir, Mersin
Syria	Latakia
Lebanon	Beirut, Tripoli
Israel	Ashdod, Haifa, Rishon LeZion, Tel Aviv
Palestinian territories	Gaza City
Egypt	Alexandria, Damietta, Port Said
Libya	Benghazi, Khoms, Misrata, Tripoli
Tunisia	Sfax, Tunis
Algeria	Algiers, Annaba, Oran
Morocco	Tétouan, Tangier

Subdivisions

According to the International Hydrographic Organization (IHO), the Mediterranean Sea is subdivided into a number of smaller waterbodies, each with their own designation (from west to east):[12]

Africa (right, on horizon) and Europe (left), as seen from Gibraltar.

- the Strait of Gibraltar;
- the Alboran Sea, between Spain and Morocco;
- the Balearic Sea, between mainland Spain and its Balearic Islands;
- the Ligurian Sea between Corsica and Liguria (Italy);
- the Tyrrhenian Sea enclosed by Sardinia, Italian peninsula and Sicily;
- the Ionian Sea between Italy, Albania and Greece;
- the Adriatic Sea between Italy, Slovenia, Croatia, Bosnia and Herzegovina, Montenegro and Albania;
- the Aegean Sea between Greece and Turkey.

Other seas

Although not recognised by the IHO treaties, there are some other seas whose names have been in common use from the ancient times, or in the present:

- the Sea of Sardinia, between Sardinia and Balearic Islands, as a part of the Balearic Sea
- the Sea of Sicily between Sicily and Tunisia,
- the Libyan Sea between Libya and Crete,
- In the Aegean Sea,
  - the Thracian Sea in its north,
  - the Myrtoan Sea between the Cyclades and the Peloponnese,
  - the Sea of Crete north of Crete,
  - the Icarian Sea between Kos and Chios
- the Cilician Sea between Turkey and Cyprus
- the Levantine Sea at the eastern end of the Mediterranean

The Torre della Pelosa at Stintino with the Sea of Sardinia behind it

Other features

Many of these smaller seas feature in local myth and folklore and derive their names from these associations. In addition to the seas, a number of gulfs and straits are also recognised:

- the Saint George Bay in Beirut, Lebanon
- the Ras Ibn Hani cape in Latakia, Syria
- the Ras al-Bassit cape in northern Syria.
- the Minet el-Beida ("White Harbour") bay near ancient Ugarit, Syria
- the Strait of Gibraltar, connects the Atlantic Ocean to the Mediterranean Sea and separates Spain from Morocco
- the Bay of Gibraltar, at the southern end of the Iberian Peninsula
- the Gulf of Corinth, an enclosed sea between the Ionian Sea and the Corinth Canal
- the Pagasetic Gulf, the gulf of Volos, south of the Thermaic Gulf, formed by the Mount Pelion peninsula
- the Saronic Gulf, the gulf of Athens, between the Corinth Canal and the Mirtoan Sea
- the Thermaic Gulf, the gulf of Thessaloniki, located in the northern Greek region of Macedonia
- the Kvarner Gulf, Croatia
- the Gulf of Lion, south of France
- the Gulf of Valencia, east of Spain
- the Strait of Messina, between Sicily and the toe of Italy
- the Gulf of Genoa, northwestern Italy
- the Gulf of Venice, northeastern Italy
- the Gulf of Trieste, northeastern Italy
- the Gulf of Taranto, southern Italy
- the Gulf of Salerno, southwestern Italy
- the Gulf of Gaeta, southwestern Italy
- the Gulf of Squillace, southern Italy
- the Strait of Otranto, between Italy and Albania
- the Gulf of Haifa, northern Israel
- the Gulf of Sidra, between Tripolitania (western Libya) and Cyrenaica (eastern Libya)

View of the Saint George Bay, and snow-capped Mount Sannine from the Corniche, Beirut

- the Strait of Sicily, between Sicily and Tunisia
- the Corsica Channel, between Corsica and Italy
- the Strait of Bonifacio, between Sardinia and Corsica
- the Gulf of İskenderun, between İskenderun and Adana (Turkey)
- the Gulf of Antalya, between west and east shores of Antalya (Turkey)
- the Bay of Kotor, in south-western Montenegro and south-eastern Croatia
- the Malta Channel, between Sicily and Malta
- the Gozo Channel, between Malta Island and Gozo

10 Major islands by area

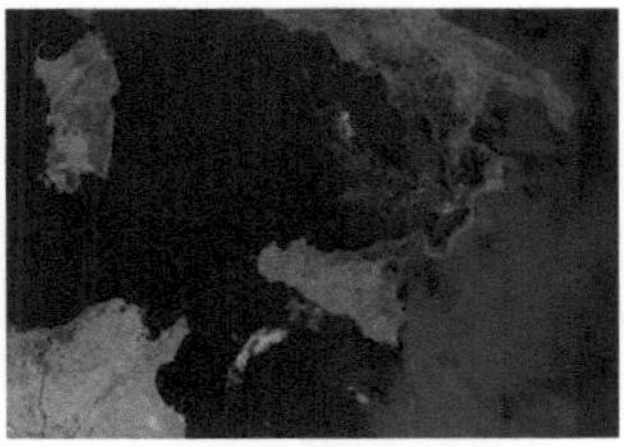

The two biggest islands of the Mediterranean: Sicily and Sardinia

Flag	Island	Area in km²	Population
	Sicily	25,460	5,048,995
	Sardinia	24,090	1,672,804
	Cyprus	9,251	1,088,503
	Corsica	8,680	299,209
	Crete	8,336	623,666
	Euboea	3,655	218.000
	Majorca	3,640	869,067
	Lesbos	1,632	90,643
	Rhodes	1,400	117,007
	Chios	842	51,936

Climate

Sea temperature

The Port of Marseille seen from L'Estaque

Sea temperature (°C)

	Jan	Feb	Mar	Apr	May	Jun	Jul	Aug	Sep	Oct	Nov	Dec
Marseille [17]	13	12	11	13	16	18	21	22	21	18	16	14
Barcelona [18]	13	13	12	14	17	20	23	25	23	20	17	15
Valencia [19]	14	13	14	15	17	21	24	26	24	21	18	15
Naples [20]	15	14	14	15	18	22	25	27	25	22	19	16
Málaga [21]	16	15	15	16	17	20	22	23	22	20	18	16
Gibraltar [22]	16	15	16	16	17	20	22	22	22	20	18	17
Athens [23]	16	15	15	16	18	21	24	24	24	21	19	18
Heraklion [24]	16	15	15	16	19	22	24	25	24	22	20	18
Malta [25]	16	16	15	16	18	21	24	26	25	23	21	18
Larnaca [26]	18	17	17	18	20	24	26	27	27	25	22	19
Limassol [27]	18	17	17	18	20	24	26	27	27	25	22	19
Antalya	17	17	17	18	21	24	27	28	27	25	22	19
Alexandria [28]	18	17	17	18	20	23	25	26	26	25	22	20
Tel Aviv [29]	18	17	17	18	21	24	26	28	27	26	23	20

Geology

A submarine karst spring, called *vrulja*, observed through sea surface rippling near Omiš

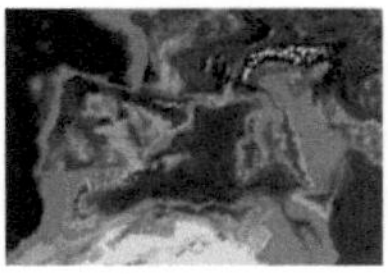

6 mya ago

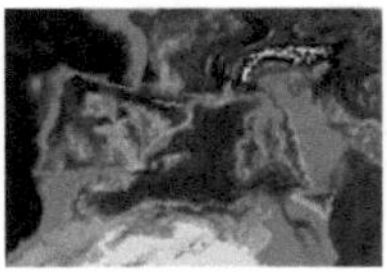

The closing of Gibraltar

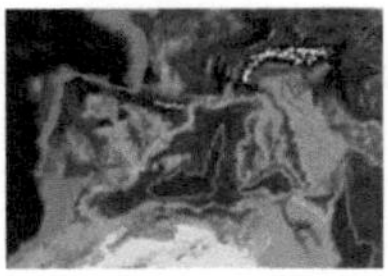

Messinian salinity crisis, 5.96 to 5.33 mya

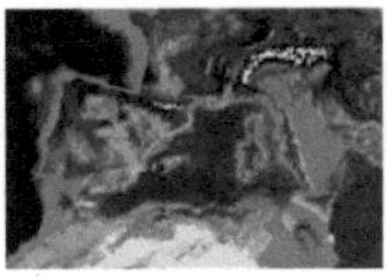

Zanclean flood

The geologic history of the Mediterranean is complex. It was involved in the tectonic break-up and then collision of the African and Eurasian plates. The Messinian Salinity Crisis occurred in the late Miocene (12 million years ago to 5 million years ago) when the Mediterranean dried up. Geologically the Mediterranean is underlain by oceanic crust.

The Mediterranean Sea has an average depth of 1500 m (**unknown operator: u'strong'** ft) and the deepest recorded point is 5267 m (**unknown operator: u'strong'** ft) in the Calypso Deep in the Ionian Sea. The coastline extends for 46000 km (**unknown operator: u'strong'** mi). A shallow submarine ridge (the Strait of Sicily) between the island of Sicily and the coast of Tunisia divides the sea in two main subregions (which in turn are divided into subdivisions), the Western Mediterranean and the Eastern Mediterranean. The Western Mediterranean covers an area of about 0.85

million km² (0.33 million mi²) and the Eastern Mediterranean about 1.65 million km² (0.64 million mi²). A characteristic of the Mediterranean Sea are submarine karst springs or *vruljas*, which mainly occur in shallow waters[30] and may also be thermal.[31]

Tectonic evolution

The geodynamic evolution of the Mediterranean Sea was provided by the convergence of European and African plates and several smaller microplates. This process was driven by the differential seafloor spreading along the Mid-Atlantic Ridge, which led to the closure of the Tethys Ocean and eventually to the Alpine orogenesis. However, the Mediterranean also hosts wide extensional basins and migrating tectonic arcs, in response to its land-locked configuration.

According to a report published by Nature in 2009, scientists think that the Mediterranean Sea was mostly filled during a time period of less than two years, in a major flood (the *Zanclean flood*) that happened approximately 5.33 million years ago, in which water poured in from the Atlantic Ocean and through the Strait of Gibraltar, at a rate three times the current flow of the Amazon River.[32]

Eastern Mediterranean

In middle Miocene times, the collision between the Arabian microplate and Eurasia led to the separation between the Tethys and the Indian oceans. This process resulted in profound changes in the oceanic circulation patterns, which shifted global climates towards colder conditions. The Hellenic arc, which has a land-locked configuration, underwent a widespread extension for the last 20 Ma due to a slab roll-back process. In addition, the Hellenic Arc experienced a rapid rotation phase during the Pleistocene, with a counterclockwise component in its eastern portion and a clockwise trend in the western segment.

Central Mediterranean

The opening of small oceanic basins of the central Mediterranean follows a trench migration and back-arc opening process that occurred during the last 30 Myr. This phase was characterised by the anticlockwise rotation of the Corsica-Sardinia block, which lasted until the Langhian (ca.16 Ma), and was in turn followed by a slab detachment along the northern African margin. Subsequently, a shift of this active extensional deformation led to the opening of the Tyrrenian basin.

Western Mediterranean

The Betic-Rif mountain belts developed during Mesozoic and Cenozoic times, as Africa and Iberia converged. Tectonic models for its evolution include: rapid motion of Alboran microplate, subduction zone and radial extensional collapse caused by convective removal of lithospheric mantle. The development of these intramontane Betic and Rif basins led to the onset of two marine gateways which were progressively closed during the late Miocene by an interplay of tectonic and glacio-eustatic processes.

Paleoenvironmental analysis

Its semi-enclosed configuration makes the oceanic gateways critical in controlling circulation and environmental evolution in the Mediterranean Sea. Water circulation patterns are driven by a number of interactive factors, such as climate and bathymetry, which can lead to precipitation of evaporites. During late Miocene times, a so-called "Messinian Salinity Crisis" (MSC hereafter) occurred, which was triggered by the closure of the Atlantic gateway. Evaporites accumulated in the Red Sea Basin (late Miocene), in the Carpatian foredeep (middle Miocene) and in the whole Mediterranean area (Messinian). An accurate age estimate of the MSC—5.96 Ma—has recently been astronomically achieved; furthermore, this event seems to have occurred synchronously. The beginning of the MSC is supposed to have been of tectonic origin; however, an astronomical control (eccentricity) might also have been involved. In the Mediterranean basin, diatomites are regularly found underneath the evaporite deposits, thus suggesting (albeit not clearly so far) a connection between their geneses.

The present-day Atlantic gateway, i.e. the Strait of Gibraltar, finds its origin in the early Pliocene. However, two other connections between the Atlantic Ocean and the Mediterranean Sea existed in the past: the Betic Corridor (southern Spain) and the Rifian Corridor (northern Morocco). The former closed during Tortonian times, thus providing a "Tortonian Salinity Crisis" well before the MSC; the latter closed about 6 Ma, allowing exchanges in the mammal fauna between Africa and Europe. Nowadays, evaporation is more relevant than the water yield supplied by riverine water and precipitation, so that salinity in the Mediterranean is higher than in the Atlantic. These conditions result in the outflow of warm saline Mediterranean deep water across Gibraltar, which is in turn counterbalanced by an inflow of a less saline surface current of cold oceanic water.

The Mediterranean was once thought to be the remnant of the Tethys Ocean. It is now known to be a structurally younger ocean basin known as Neotethys. The Neotethys formed during the Late Triassic and Early Jurassic rifting of the African and Eurasian plates.

Paleoclimate

Because of its latitudinal position and its land-locked configuration, the Mediterranean is especially sensitive to astronomically induced climatic variations, which are well documented in its sedimentary record. Since the Mediterranean is involved in the deposition of eolian dust from the Sahara during dry periods, whereas riverine detrital input prevails during wet ones, the Mediterranean marine sapropel-bearing sequences provide high-resolution climatic information. These data have been employed in reconstructing astronomically calibrated time scales for the last 9 Ma of the Earth's history, helping to constrain the time of past Geomagnetic Reversals.[33] Furthermore, the exceptional accuracy of these paleoclimatic records have improved our knowledge of the Earth's orbital variations in the past.

Ecology and biota

As a result of the drying of the sea during the Messinian salinity crisis,[34] the marine biota of the Mediterranean are derived primarily from the Atlantic Ocean. The North Atlantic is considerably colder and more nutrient-rich than the Mediterranean, and the marine life of the Mediterranean has had to adapt to its differing conditions in the five million years since the basin was reflooded.

The Alboran Sea is a transition zone between the two seas, containing a mix of Mediterranean and Atlantic species. The Alboran Sea has the largest population of Bottlenose Dolphins in the western Mediterranean, is home to the last population of harbour porpoises in the Mediterranean, and is the most important feeding grounds for Loggerhead Sea Turtles in Europe. The Alboran sea also hosts important commercial fisheries, including sardines and swordfish. In 2003, the World Wildlife Fund raised concerns about the widespread drift net fishing endangering populations of dolphins, turtles, and other marine animals.

Environmental threats

Biodiversity

Invasive species

The Reticulate whipray is one of the species that colonised the eastern Mediterranean through the Suez Canal.

The opening of the Suez Canal in 1869 created the first salt-water passage between the Mediterranean and Red Sea. The Red Sea is higher than the Eastern Mediterranean, so the canal serves as a tidal strait that pours Red Sea water into the Mediterranean. The Bitter Lakes, which are hyper-saline natural lakes that form part of the canal, blocked the migration of Red Sea species into the Mediterranean for many decades, but as the salinity of the lakes gradually equalized with that of the Red Sea, the barrier to migration was removed, and plants and animals from the Red Sea have begun to colonise the Eastern Mediterranean. The Red Sea is generally saltier and more nutrient-poor than the Atlantic, so the Red Sea species have advantages over Atlantic species in the salty and nutrient-poor Eastern Mediterranean. Accordingly, Red Sea species invade the Mediterranean biota, and not vice versa; this phenomenon is known as the Lessepsian migration (after Ferdinand de Lesseps, the French engineer) or Erythrean invasion. The construction of the Aswan High Dam across the Nile River in the 1960s reduced the inflow of freshwater and nutrient-rich silt from the Nile into the Eastern Mediterranean, making conditions there even more like the Red Sea and worsening the impact of the invasive species.

Invasive species have become a major component of the Mediterranean ecosystem and have serious impacts on the Mediterranean ecology, endangering many local and endemic Mediterranean species. A first look at some groups of exotic species show that more than 70% of the non-indigenous decapods and about 63% of the exotic fishes occurring in the Mediterranean are of Indo Pacific origin,[35] **introduced** into the Mediterranean through the Suez Canal. This makes the Canal as the first pathway of arrival of "alien" species into the Mediterranean. The impacts of some lessepsian species have proven to be considerable mainly in the Levantine basin of the Mediterranean, where they are replacing native species and becoming a "familiar sight".

According to the International Union for Conservation of Nature definition, as well as Convention on Biological Diversity(CBD) and Ramsar Convention terminologies, they are alien species, as they are non-native (non-indigenous) to the Mediterranean Sea, and they are outside their normal area of distribution which is the Indo-Pacific region. When these species succeed in establishing populations in the Mediterranean sea, compete with and begin to replace native species they are "Alien Invasive Species", as they are an agent of change and a threat to the native biodiversity. Depending on their impact, Lessepsian migrants are either alien or alien invasive species. In the context of CBD, "introduction" refers to the movement by human agency, indirect or direct, of an alien species outside of its natural range (past or present). The Suez Canal, being an artificial (man made) canal, is a human agency. Lessepsian migrants are therefore "introduced" species (indirect, and unintentional). Whatever wording is chosen, they represent a threat to the native Mediterranean biodiversity, because they are non-indigenous to this sea. In recent years, the Egyptian government's announcement of its intentions to deepen and widen the canal have raised concerns from marine biologists, fearing that such an act will only worsen the invasion of Red Sea species into the Mediterranean, facilitating the crossing of the canal for yet additional species.[36]

Arrival of new tropical Atlantic species

In recent decades, the arrival of exotic species from the tropical Atlantic has become a noticeable feature. Whether this reflects an expansion of the natural area of these species that now enter the Mediterranean through the Gibraltar strait, because of a warming trend of the water caused by Global Warming; or an extension of the maritime traffic; or is simply the result of a more intense scientific investigation, is still an open question. While not as intense as the "lessepsian" movement, the process deserves to be studied and monitored.

Sea-level rise

The overall level of the Mediterranean could rise between 3–61 cm as a result of the effects of climate change.[37] This could have adverse effects on populations across the Mediterranean:

- Rising sea levels will submerge parts of Malta. Rising sea levels will also mean rising salt water levels in Malta's groundwater supply and reduce the availability of drinking water.[38]
- A 30 cm rise in sea level would flood 200 square kilometres of the Nile Delta, displacing over 500,000 Egyptians.[39]

Coastal ecosystems also appear to be threatened by sea level rise, especially enclosed seas such as the Baltic, the Mediterranean and the Black Sea. These seas have only small and primarily east-west orientated movement corridors, which may restrict northward displacement of organisms in these areas.[40] Sea level rise for the next century (2100) could be between 30 cm (**unknown operator: u'strong'** in) and 100 cm (**unknown operator: u'strong'** in) and temperature shifts of a mere 0.05-0.1°C in the deep sea are sufficient to induce significant changes in species richness and functional diversity.[41]

Pollution

Pollution in this region has been extremely high in recent years. The United Nations Environment Programme has estimated that 650000000 t (**unknown operator: u'strong'** short tons) of sewage, 129000 t (**unknown operator: u'strong'** short tons) of mineral oil, 60000 t (**unknown operator: u'strong'** short tons) of mercury, 3800 t (**unknown operator: u'strong'** short tons) of lead and 36000 t (**unknown operator: u'strong'** short tons) of phosphates are dumped into the Mediterranean each year.[42] The Barcelona Convention aims to 'reduce pollution in the Mediterranean Sea and protect and improve the marine environment in the area, thereby contributing to its sustainable development.'[43] Many marine species have been almost wiped out because of the sea's pollution. One of them is the Mediterranean Monk Seal which is considered to be among the world's most endangered marine mammals.[44]

The Mediterranean is also plagued by marine debris. A 1994 study of the seabed using trawl nets around the coasts of Spain, France and Italy reported a particularly high mean concentration of debris; an average of 1,935 items per km². Plastic debris accounted for 76%, of which 94% was plastic bags.[45]

Shipping

Some of the world's busiest shipping routes are in the Mediterranean Sea. It is estimated that approximately 220,000 merchant vessels of more than 100 tonnes cross the Mediterranean Sea each year — about one third of the world's total merchant shipping. These ships often carry hazardous cargo, which if lost would result in severe damage to the marine environment.

The discharge of chemical tank washings and oily wastes also represent a significant source of marine pollution. The Mediterranean Sea constitutes 0.7% of the global water surface and yet receives seventeen percent of global marine oil pollution. It is estimated that every year between 100000 t (**unknown operator: u'strong'** long tons) and 150000 t (**unknown operator: u'strong'** long tons) of crude oil are deliberately released into the sea from shipping activities.

Approximately 370000000 t (**unknown operator: u'strong'** long tons) of oil are transported annually in the Mediterranean Sea (more than 20% of the world total), with around 250-300 oil tankers crossing the Sea every day. Accidental oil spills happen frequently with an average of 10 spills per year. A major oil spill could occur at any time in any part of the Mediterranean.[41]

Tourism

Barceloneta beach of Barcelona during the tourist high season.

With a unique combination of pleasant climate, beautiful coastline, rich history and diverse culture the Mediterranean region is the most popular tourist destination in the world — attracting approximately one third of the world's international tourists.

Tourism is one of the most important sources of income for many Mediterranean countries. It also supports small communities in coastal areas and islands by providing alternative sources of income far from urban centres. However, tourism has also played major role in the degradation of the coastal and marine environment. Rapid development has been encouraged by Mediterranean governments to support the large numbers of tourists visiting the region each year. But this has caused serious disturbance to marine habitats such as erosion and pollution in many places along the Mediterranean coasts.

Tourism often concentrates in areas of high natural wealth, causing a serious threat to the habitats of endangered Mediterranean species such as sea turtles and monk seals. It is ironic that tourism in this region is destroying the foundations of its own existence. And it is inevitable that the tourists will leave the Mediterranean as it becomes more depleted of its natural beauty.[41]

Overfishing

Fish stock levels in the Mediterranean Sea are alarmingly low. The European Environment Agency says that over 65% of all fish stocks in the region are outside safe biological limits and the United Nations Food and Agriculture Organisation, that some of the most important fisheries — such as albacore and bluefin tuna, hake, marlin, swordfish, red mullet and sea bream — are threatened.

There are clear indications that catch size and quality have declined, often dramatically, and in many areas larger and longer-lived species have disappeared entirely from commercial catches.

Large open water fish like tuna have been a shared fisheries resource for thousands of years but the stocks are now dangerously low. In 1999, Greenpeace published a report revealing that the amount of bluefin tuna in the Mediterranean had decreased by over 80% in the previous 20 years and government scientists warn that without immediate action the stock will collapse.

Aquaculture

Aquaculture is expanding rapidly — often without proper environmental assessment — and currently accounts for 30% of the fish protein consumed worldwide. The industry claims that farmed seafood lessens the pressure on wild fish stocks, yet many of the farmed species are carnivorous, consuming up to five times their weight in wild fish.

Aquaculture in western Greece

Mediterranean coastal areas are already over exposed to human influence, with pristine areas becoming ever scarcer. The aquaculture sector adds to this pressure, requiring areas of high water quality to set up farms. The installation of fish farms close to vulnerable and important habitats such as seagrass meadows is particularly concerning.

Gallery

Old city of Ibiza Town, Spain

The beach of la Courtade in the Îles d'Hyères, France

Sardinia's east coast, Italy

Coast of Żurrieq, Malta

Panoramic view of Cavtat, Croatia

A view of Sveti Stefan, Montenegro

Ksamil beaches, Albania

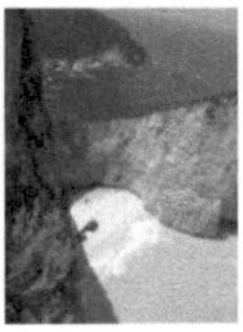

Navagio (shipwreck) bay in Zakynthos island, Greece

Burj Islam Beach, Latakia, Syria

A view of Raouché off the coast of Beirut, Lebanon

North Israeli coastal plain, Israel

Coast of Alexandria, A view From Bibliotheca Alexandrina, Egypt

a beach in Benghazi, Libya

Beach of Hammamet, Tunisia

Les Aiguades near Béjaïa, Algeria

El Jebha, a port town in Morocco

Europa Point, Gibraltar United Kingdom

See also

- Babelmed, the site of the Mediterranean cultures
- Euro-Mediterranean Parliamentary Assembly
- Familial Mediterranean fever
- History of the Mediterranean region
- Holy League (Mediterranean)
- List of islands in the Mediterranean
- List of Mediterranean countries
- Mediterranean Basin
- Mediterranean climate
- Mediterranean diet
- Mediterranean forests, woodlands, and scrub
- Mediterranean Games
- Mediterranean race
- Mediterranean sea (oceanography)
- Mediterranean Union
- Piri Reis, early Mediterranean maps
- Seto Inland Sea, which is sometimes named the Japanese Mediterranean Sea
- Suez Canal

References

[1] http://www.unece.org/fileadmin/DAM/env/water/blanks/assessment/black.pdf

[2] Pinet, Paul R. (2008). *Invitation to Oceanography* (http://books.google.com/books?id=6TCm8Xy-sLUC&pg=PA220&lpg=PA220). Jones & Barlett Learning. p. 220. ISBN 0-7637-5993-7. .

[3] "Microsoft Word - ext_abstr_East_sea_workshop_TLM.doc" (http://www.ifremer.fr/lobtln/OTHER/ext_abstr_East_Sea_workshop_TLM.pdf) (PDF). . Retrieved 2010-04-23.

[4] "Researchers predict Mediterranean Sea level rise - Headlines - Research – European Commission" (http://ec.europa.eu/research/headlines/news/article_09_03_19_en.html). Ec.europa.eu. 2009-03-19. . Retrieved 2010-04-23.

[5] Georg Wilhelm Friedrich Hegel *The Philosophy of History*, p. 87, Dover Publications Inc., 1956 ISBN 0-486-20112-0; 1st ed. 1899

[6] entry μεσόγαιος (http://old.perseus.tufts.edu/cgi-bin/ptext?doc=Perseus:text:1999.04.0057:entry=#66422) at Liddell & Scott

[7] David Abulafia, *The Great Sea: A Human History of the Mediterranean* (Oxford University Press; 2011)

[8] Rappoport, S. (Doctor of Philosophy, Basel). History of Egypt (undated, early 20th century), Volume 12, Part B, Chapter V: "The Waterways of Egypt", pages 248-257. London: The Grolier Society.

[9] C.I. Gable - Constantinople Falls to the Ottoman Turks (http://www.boglewood.com/timeline/ottomans.html) - *Boglewood Timeline* - 1998 - Retrieved 3 September 2011.

[10] "History of the Ottoman Empire, an Islamic Nation where Jews Lived" (http://www.sephardicstudies.org/ottoemp.html) - *Sephardic Studies and Culture* - Retrieved 3 September 20111.

[11] Robert Guisepi - The Ottomans: From Frontier Warriors To Empire Builders (http://history-world.org/ottoman empire.htm) - 1992 - *History World International* - Retrieved 3 September 2011.

[12] "Limits of Oceans and Seas, 3rd edition" (http://www.iho-ohi.net/iho_pubs/standard/S-23/S23_1953.pdf). International Hydrographic Organization. 1953. . Retrieved 7 February 2010.

[13] Pinet, Paul R. (1996), *Invitation to Oceanography* (3rd ed.), St Paul, Minnesota: West Publishing Co., p. 202, ISBN 0-314-06339-0

[14] Pinet 1996, p. 206.
[15] Pinet 1996, pp. 206–207.
[16] Pinet 1996, p. 207.
[17] http://www.weather2travel.com/climate-guides/france/marseille.php
[18] http://www.weather2travel.com/climate-guides/spain/barcelona.php
[19] http://www.weather2travel.com/climate-guides/spain/valencia.php
[20] http://www.weather2travel.com/climate-guides/italy/neapolitan-riviera/naples.php
[21] http://www.weather2travel.com/climate-guides/spain/costa-del-sol/malaga.php
[22] http://www.weather2travel.com/climate-guides/gibraltar/gibraltar.php
[23] http://www.weather2travel.com/climate-guides/greece/athens.php
[24] http://www.weather2travel.com/climate-guides/greece/crete/iraklion.php
[25] http://www.weather2travel.com/climate-guides/malta/valletta.php
[26] http://www.weather2travel.com/climate-guides/cyprus/larnaca.php
[27] http://www.weather2travel.com/climate-guides/cyprus/limassol.php
[28] http://www.weather2travel.com/climate-guides/egypt/alexandria.php
[29] http://www.weather2travel.com/climate-guides/israel/tel-aviv.php
[30] Elmer LaMoreaux, Philip (2001). "Geologic/Hydrogeologic Setting and Classification of Springs" (http://books.google.si/books?id=sjEoBmfUka0C&pg=PA58#v=onepage&q&f=false). *Springs and Bottled Waters of the World: Ancient History, Source, Occurence, Quality and Use*. Springer. p. 57. ISBN 978-3-540-61841-6. .
[31] Žumer, Jože (2004). "Odkritje podmorskih termalnih izvirov [Discovery of submarine thermal springs]" (http://zgs.zrc-sazu.si/Portals/8/Geografski_obzornik/go_2004_2.pdf) (in Slovene). *Geografski obzornik* (Association of the Geographical Societies of Slovenia) **51** (2): 11–17. ISSN 0016-7274. . (Slovene)
[32] Mediterranean Sea filled in less than two years: study (http://www.breitbart.com/article.php?id=CNG.4d3eaceb94e76f30f9e2b75555025c27.3d1&show_article=1), AFP, December 9, 2009
[33] FJ, Hilgen. Astronomical calibration of Gauss to Matuyama sapropels in the Mediterranean and implication for the Geomagnetic Polarity Time Scale, 104 (1991) 226-244 *Earth and Planetary Science Letters*, 1991. (http://igitur-archive.library.uu.nl/geo/2007-0119-200857/hilgen_91_astronomical.pdf)
[34] Hsu K.J., "When the Mediterranean Dried Up" *Scientific American*, Vol. **227**, December 1972, p32
[35] "IUCN Guidelines for the Prevention of Biodiversity Loss Caused by Alien Invasive Species" (http://www.issg.org/infpaper_invasive.pdf). International Union for Conservation of Nature. 2000. . Retrieved 2009-08-11.
[36] Galil, B.S. and Zenetos, A. (2002). A sea change: exotics in the eastern Mediterranean Sea, in: Leppäkoski, E. et al. (2002). *Invasive aquatic species of Europe: distribution, impacts and management*. pp. 325-336.
[37] "Mediterranean Sea Level Could Rise By Over Two Feet, Global Models Predict" (http://www.sciencedaily.com/releases/2009/03/090303084057.htm). Science Daily. 2009-03-03. .
[38] "Briny future for vulnerable Malta" (http://news.bbc.co.uk/2/hi/science/nature/6525069.stm). BBC News. 2007-04-04. .
[39] "Egypt fertile Nile Delta falls prey to climate change" (http://news.egypt.com/en/201001288902/news/-egypt-news/egypt-fertile-nile-delta-falls-prey-to-climate-change.html). 2010-01-28. .
[40] Nicholls, R.J.; Klein,R.J.T. (2005). Climate change and coastal management on Europe's coast, in: Vermaat, J.E. et al. (Ed.) (2005). Managing European coasts: past, present and future. pp. 199-226.
[41] "Other threats in the Mediterranean | Greenpeace International" (http://www.greenpeace.org/international/campaigns/oceans/marine-reserves/the-mediterranean/mediterranean-other-threats). Greenpeace.org. . Retrieved 2010-04-23.
[42] "Pollution in the Mediterranean Sea. Environmental issues" (http://www.explorecrete.com/nature/mediterranean.html). Explorecrete.com. . Retrieved 2010-04-23.
[43] "EUROPA" (http://europa.eu/scadplus/leg/en/lvb/l28084.htm). Europa.eu. . Retrieved 2010-04-23.
[44] "Mediterranean Monk Seal Fact Files: Overview" (http://www.monachus-guardian.org/factfiles/medit01.htm). Monachus-guardian.org. 1978-05-05. . Retrieved 2010-04-23.
[45] publications/docs/anl_oview.pdf "Marine Litter: An analytical overview" (http://www.unep.org/regionalseas/marinelitter). United Nations Environment Programme. 2005. publications/docs/anl_oview.pdf. Retrieved 2008-08-01.

Philip Mansel, *Levant: Splendour and Catastrophe on the Mediterranean*, London, John Murray, 11 November 2010, hardback, 480 pages, ISBN 978-0-7195-6707-0, New Haven, Yale University Press, 24 May 2011, hardback, 470 pages, ISBN 978-0-300-17264-5

External links

- Greenpeace campaign "Defending Our Mediterranean": Threats, Solutions and Photo Petition (http://www.greenpeace.org/lebanon/en)
- Planblue - Environment and Development in the Mediterranean Region (http://www.planbleu.org/indexUK.html)
- Finding Nino (http://www.findingnino.com.au) By Marc Llewellyn - winner of the ASTW Travel Book of the Year Award, about the Mediterranean, finding oneself, and becoming a peasant farmer.
- Mediterranean Sea Microorganisms: 180+ images of Foraminifera (http://www.foraminifera.eu/querydb.php?area=Mediterranean+Sea&aktion=suche)

Harland_and_Wolff

Type	Private Limited Company
Industry	Civil Engineering Marine Engineering Shipbuilding Offshore construction
Founded	1861 Nationalised 1977 Privatised 1989
Headquarters	Belfast, Northern Ireland, UK
Key people	Edward Harland Gustav Wilhelm Wolff William Pirrie, 1st Viscount Pirrie Thomas Andrews Sir John Parker Fred Olsen
Owner(s)	Fred. Olsen Energy
Employees	500
Website	www.harland-wolff.com [1]

Harland and Wolff Heavy Industries is a Northern Irish heavy industrial company, specializing in shipbuilding and offshore construction, located in Belfast, Northern Ireland.

The shipyard has built many ships; among the more famous are the White Star trio *Olympic*, *Titanic* and *Britannic*, the Royal Navy's HMS *Belfast*, Royal Mail's *Andes*, Shaw Savill's *Southern Cross* and P&O's *Canberra*. The company's official history, "Shipbuilders to the World" was published in 1986.[2]

As of 2011, the expanding offshore wind power industry has taken centre stage and 75% of the company's work is based on offshore renewable energy.[3]

Early history

Harland and Wolff was formed in 1861 by Edward James Harland (1831–1895) and Hamburg-born Gustav Wilhelm Wolff (1834–1913, in the UK from age 14). In 1858 Harland, then general manager, bought the small shipyard on *Queen's Island* from his employer Robert Hickson.

Workers leaving the shipyard at Queens Road in early 1911. The RMS *Titanic* can be seen in the background, underneath the Arrol gantry.

After buying Hickson's shipyard, Harland made his assistant Wolff a partner in the company. Wolff was the nephew of Gustav Schwabe, Hamburg, who was heavily invested in the Bibby Line, and the first three ships that the newly incorporated shipyard built were for that line. Harland made a success of the business through several innovations, notably replacing the wooden upper decks with iron ones which increased the strength of the ships; and giving the hulls a flatter bottom and squarer cross section, which increased their capacity.

Statue of Edward James Harland in the grounds of Belfast City Hall

When Harland died in 1895, William James Pirrie became the chairman of the company until his death in 1924. Thomas Andrews also became the general manager and head of the draughting department in 1907. It was during this period that the company built the RMS *Olympic* and her sister-ships RMS *Titanic* and HMHS *Britannic* between 1909 and 1914, commissioning Sir William Arrol & Co. to construct a massive twin gantry and slipway structure for the project.

In 1912, the company acquired another shipyard at Govan in Glasgow, Scotland. It bought the former London & Glasgow Engineering & Iron Shipbuilding Co's Middleton and Govan New shipyards in Govan and Mackie & Thomson's Govan Old yard, which had been owned by William Beardmore and Company. The three neighbouring yards were amalgamated and redeveloped to provide a total of seven building berths, a fitting-out basin and extensive workshops. Harland & Wolff specialised in building tankers and cargo ships at Govan. The nearby shipyard of A. & J. Inglis was also purchased by Harland and Wolff in 1919, along with a stake in the company's primary steel supplier, David Colville & Sons. Harland & Wolff also established shipyards at Bootle in Liverpool, North Woolwich in London and Southampton. These shipyards were all eventually closed from the early 1960s however, when the company opted to consolidate its operations in Belfast.

The Harland & Wolff Belfast Drawing Offices during the early 20th century

The war years

A burner operating at night on the deck of a ship at Harland and Wolff's Liverpool yard (27 October 1944).

During World War I, Harland and Wolff built monitors and cruisers, including the 15-inch gun armed "large light cruiser" HMS *Glorious*. In 1918, the company opened a new shipyard on the eastern side of the Musgrave Channel which was named the East Yard. This yard specialised in mass-produced ships of standard design developed during the First World War.

The company started an aircraft manufacturing subsidiary with Short Brothers, called **Short and Harland Limited** in 1936. Its first order was for 189 Handley Page Hereford bombers built under license from Handley Page for the Royal Air Force. During the Second World War, this factory built Short Stirling bombers as the Hereford was removed from service.

The shipyard was busy during World War II, building 6 aircraft carriers, 2 cruisers (including HMS *Belfast*) and 131 other naval ships; and repairing over 22,000 vessels. It also manufactured tanks and artillery components. It was during this period that the company's workforce peaked at around 35,000 people. However, many of the vessels built during this era were commissioned right at the end of World War II, as Harland and Wolff were focused on ship repair during the first three years of the war. The yard on Queen's Island was heavily bombed by the Luftwaffe in April and May 1941 causing considerable damage to the shipbuilding facilities and destroying the aircraft factory.

Post-war period

With the rise of the jet-powered airliner in the late 1950s, the demand for ocean liners declined. This, coupled with competition from Japan, led to difficulties for the British shipbuilding industry. The last liner that the company launched was the MV *Arlanza* for Royal Mail Line in 1960, whilst the last liner completed was the SS *Canberra* for P&O in 1961.

In the 1960s, notable achievements for the yard included the tanker Myrina which was the first supertanker built in the UK, and the largest vessel ever launched down a slipway (September 1967). In the same period the yard also built the semisubmersible drilling rig *Sea Quest* which, due to its three-legged design, was launched down three parallel slipways. This was a first and only time this was ever done.

In the mid-1960s, the British government started advancing loans and subsidies to British shipyards to preserve jobs. Some of this money was used to finance the modernisation of the yard, allowing it to build the much larger post-war merchant ships including one of 333,000 tonnes. However continuing problems led to the company's nationalisation, though not as part of British Shipbuilders, in 1977.

The company was bought from the British government in 1989 in a management/employee buy-out in partnership with the Norwegian shipping magnate Fred Olsen; leading to a new company called **Harland and Wolff Holdings Plc**. By this time, the number of people employed by the company had fallen to around 3,000.

For the next few years, Harland and Wolff specialised in building standard Suezmax oil tankers, and has continued to concentrate on vessels for the offshore oil and gas industry. It has made some forays outside of this market. The company bid unsuccessfully tendered against Chantiers de l'Atlantique for the construction of Cunard line's new *Queen Mary 2*.[4]

In the late 1990s, the yard was part of the then British Aerospace's team for the Royal Navy's Future Carrier (CVF) programme. It was envisaged that the ship would be assembled at the Harland & Wolff dry-dock in Belfast. In 1999 BAe merged with Marconi Electronic Systems. The new company, BAE Systems Marine, included the former Marconi shipyards on the Clyde and at Barrow-in-Furness thus rendering H&W's involvement surplus to requirements.

Restructuring and current operations

The Samson and Goliath gantry cranes have become city landmarks.

Faced with competitive pressures (especially as regards shipbuilding), Harland and Wolff sought to shift and broaden their portfolio, focusing less on shipbuilding and more on design and structural engineering, as well as ship repair, offshore construction projects and competing for other projects to do with metal engineering and construction. This led to Harland and Wolff constructing a series of bridges in Britain and also in the Republic of Ireland, such as the James Joyce Bridge and the restoration of Dublin's Ha'penny Bridge, building on the success of its first foray into the civil engineering sector with the construction of the Foyle Bridge in the 1980s.

Harland & Wolff is now a leading offshore fabrication and ship repair yard.

Harland and Wolffs last shipbuilding project (to date) was the MV *Anvil Point*, one of six near identical Point class sealift ships built for use by the Ministry of Defence. The ship, built under licence from German shipbuilders Flensburger Schiffbau-Gesellschaft, was launched in 2003.

In recent years the company has indeed seen its ship-related workload increase slightly. Whilst Harland and Wolff has no involvement in any shipbuilding projects for the foreseeable future, the company is increasingly involved in overhaul, re-fitting and ship repair, as well as the construction and repair of off-shore equipment such as oil platforms. On 1 February 2011 it was announced that Harland and Wolff had won the contract to refurbish the SS *Nomadic*, effectively rekindling its nearly 150-year association with the White Star Line. Structural steel work on the ship began on February 10, 2011 and has been completed in time for the 2012 Belfast Titanic Festival. In July 2012 Harland and Wolff will carry out the dry docking and service of the Husky Oil SeaRose FPSO (Floating Production, Storage and Offloading) vessel.

MS *Isle of Inishmore* and HSC *Jonathan Swift* undergoing refit at Harland & Wolff in 2008.

Belfast's skyline is still dominated today by Harland and Wolff's famous twin Gantry cranes, Samson and Goliath, built in 1974 and 1969 respectively. There is also speculation about a resurgence in the prosperity of the shipyard thanks to the company's diversification into emerging technologies, particularly in renewable energy development, such as offshore wind turbine and tidal power construction, which may provide an opportunity to further improve the company's fortunes in the long term. For example, the United Kingdom planned to build 7,500 new offshore wind turbines between 2008 and 2020,[5] creating great demand for heavy assembly work. Unlike land-based wind turbines, where assembly occurs on site, offshore wind turbines have part of their assembly done in a shipyard, and then construction barges transport the tower sections, rotors, and nacelles to the site for final erection and assembly. As a result of this, in late 2007, the 'Goliath' gantry crane was re-commissioned, having been moth-balled in 2003 due to the lack of heavy-lifting work at the yard.

In June 2008, assembly work at the Belfast yard was underway on 60 Vestas V90-3MW wind turbines for the Robin Rigg Wind Farm.[6] This was the second offshore wind farm assembled by the company for Vestas having completed the logistics for the Barrow Offshore Wind Farm in 2006. In August 2011 Harland and Wolff completed the logistics for the Ormonde Wind Farm which consisted of 30 REpower 5MW turbines.

In March 2008, the construction of the world's first commercial tidal stream turbine, for Marine Current Turbines, was completed at the Belfast yard. The installation of the 1.2MW SeaGen Tidal System was begun in Strangford Lough in April 2008.[7]

In July 2010, Harland and Wolff secured a contract to make a prototype tidal energy turbine for Scotrenewables Ltd.[8] Manufacture of the SR250 device was completed in May 2011 and has been undergoing testing in Orkney since.

As of April 2012, the booming offshore wind power industry has taken centre stage. Harland and Wolff are currently working on three innovative meteorological mast foundations for the Dogger Bank and Firth of Forth offshore wind farms, as well as putting the finishing touches to two Siemens substations for the Gwynt y Môr offshore wind farm. Seventy-five per cent of the company's work is based on offshore renewable energy. Harland & Wolff is one of many UK and international companies profiting from the emergence of UK wind- and marine-generated electricity, which is attracting significant inward investment.[3]

Archives

The archives relating to the Harland and Wolff Heavy Industries are maintained by the Archives of the University of Glasgow (GUAS). A collection of Harland and Wolff papers are held at Public Record Office of Northern Ireland (PRONI).[9] Their "Introduction Harland and Wolff Papers" [10] issued 2007, notes "The Harland & Wolff archive in PRONI comprises c.2,000 files, c.200 volumes and c.16,000 documents, 1861-1987, documenting most aspects of the history of Belfast's famous shipbuilding firm". A further major archive is held at the Ulster Folk and Transport Museum (UFTM). This has a photographic collection and a ships' plans collection (i.e., technical drawings). While over 2,000 photographs in the UFTM Welch Collection [11] may be accessed, the plans collection is not advertised on the UFTM website. According to PRONI (2007:10),[12] "The UFTM's collection of ships' plans are not available to the public in any shape or form at the moment. There is no copy service". Selected early ship's plans (dating from 1860 to 1882) are reproduced in a pictorial book by McCluskie (1998).[13]

List of ships built

Ships built by Harland and Wolff include:[14] [15]

> *This transport-related list is incomplete; you can help by expanding it* [16].

.

SS Venetian - the first steamer built by Harland & Wolff 1860 - a plaque on the William James Pirrie monument, grounds of Belfast City Hall.

- SS *Venetian*, completed 1860 for Bibby Line
- RMS *Atlantic*, built 1870, maiden voyage 8 June 1871, sank 1 April 1873
- RMS *Oceanic*, launched on 27 August 1870, maiden voyage 2 March 1871. The White Star Line's first liner. H+W also built 88 other ships for White Star
- SS *Baltic*, later *Veendam*, sank 6 February 1898
- SS *Adriatic*, launched on 17 October 1871, maiden voyage 11 April 1872
- SS *Celtic*, launched on 18 June 1872, maiden voyage 24 October 1872
- SS *Britannic*, maiden voyage 25 June 1874, scrapped 1903
- SS *British Crown*, launched 1879, maiden voyage 15 October 1879, renamed *Amsterdam* 1887
- SS *British Queen*, launched 1880, maiden voyage 31 January 1881, renamed *Onega* 1915, sunk by torpedo 1918
- SS *Germanic*, launched 1874, scrapped 1950
- SS *Coptic*, launched 10 August 1881, maiden voyage 16 November 1881
- SS *Doric*, launched 1883, 1906 renamed *Asia*
- SS *Ionic*, launched 1884, scrapped 1908
- SS *Gaika*, launched 1897, scrapped 1929
- SS *Majestic*, launched 1889, scrapped 1914
- SS *Michigan*, Launched April 19, 1890, Maiden Voyage on June 24, 1890; Renamed *Kilpatrick*, *Acropolis*, *Washington*, *Great Canton*; Scrapped in Italy 1924.
- SS *Mississippi*, Launched August 29, 1890, Maiden Voyage on October 28, 1890; Renamed USAT *Buford*; Scrapped in Japan 1929.
- SS *Massachusetts*, Launched on December 17, 1891; Maiden Voyage on April 24, 1892; Renamed the Sheridan; Scrapped in October 1923.
- SS *Manitoba*, Launched on January 7, 1892; Maiden Voyage on April 15, 1892; Renamed the Logan, Candler; Scrapped in 1926.
- SS *Mohawk*, Launched in 1892; Maiden Voyage on April 1892; Renamed the Grant, Chinouk; Scrapped in 1946.
- SS *Mobile*, Launched on January 20, 1893; Renamed the Sherman, Calawaii; Scrapped in Japan in 1933.

- SS *Gothic*, maiden voyage 28 December 1893, scrapped 1926
- SS *Minnewaska*, Launched in 1894, previously named *Persia*, later renamed USAT *Thomas*; Scrapped in 1929.
- SS *Armenian*, launched 25 November 1895, sunk by torpedo 1915
- SS *Canada*, launched on 14 May 1896, maiden voyage 1 October 1896
- SS *Cymric*, launched 1898, sunk April 13, 1916
- SS *New England*, launched April 7, 1898, maiden voyage June 30, 1898, renamed *Romanic* November 1913, scrapped 1922
- RMS *Oceanic*, launched on 14 January 1899
- SS *Commonwealth*, launched on 31 May 1900, maiden voyage 4 October 1900. Became *Canopic* (1904)
- SS *Minnehaha*, launched on March 31, 1900; maiden voyage on July 7, 1900; torpedoed and sunk by U-Boat U 48 on September 7, 1917.
- RMS *Celtic*, launched on 4 April 1901, maiden voyage 26 July 1901. Wrecked 10 December 1928, Roche's Point, Cobh, Ireland
- RMS *Walmer Castle*, launched on 6 July 1901
- SS *Athenic*, launched on 17 August 1901, maiden voyage 13 February 1902
- USS *Artemis*, launched 15 August 1902 as *Iowa*, later *Bohemia* and *Empire Bittern*, sunk 1944
- SS *Cedric*, launched on 21 August 1902, maiden voyage 11 February 1903
- WSL *Corinthic* launched 1902, scrapped 1932
- SS *Ionic*, launched on 22 May 1903, scrapped in Japan, 1936
- RMS *Baltic*, launched 21 November 1903, scrapped in Japan in 1933
- RMS *Kenilworth Castle*, launched on 15 December 1903, completed May 1904
- SS *Mamari*, maiden voyage 15 December 1904
- RMS *Aragon*, launched on 23 February 1905, maiden voyage 14 July 1905
- USS *America*, launched 20 April 1905 as SS *Amerika, later USAT* America *and USAT* Edmund B. Alexander, *scrapped 1957*
- TSS *Nieuw Amsterdam*, launched September 28, 1905; Maiden Voyage April 7, 1906; Scrapped in Japan February 1932.
- SS *Rohilla*, completed 1906; lost off Whitby 30 October 1914.
- USS *Republic*, launched in 1907
- SS *Megantic*, launched 1908, scrapped in Japan, 1933
- SS *Laurentic*, launched 1908, sunk by mines January 1917
- RMS *Edinburgh Castle*, launched on 27 January 1910, completed 28 April 1910, maiden voyage May 1910
- SS *Pakeha*, launched on 26 May 1910, completed 20 August 1910
- RMS *Olympic*, launched 20 October 1910, maiden voyage 14 June 1911
- SS *Nomadic*, launched 25 April 1911, tender to RMS *Titanic* and RMS *Olympic*, under restoration 2011
- SS *Traffic*, launched 27 April 1911, sunk 1941
- RMS *Titanic*, *Olympic* class launched on 31 May 1911, maiden voyage 10 April 1912, sunk April 1912
- SS *Zealandic*, launched on 29 June 1911, maiden voyage 30 October 1911, sunk 1941
- RMS *Arlanza*, launched 23 November 1911, completed September 1912
- SS *Ceramic*, launched on 11 December 1912, completed 5 July 1913
- SS *Pittsburgh*, launched 1913, entered service 1922, renamed *Pennland*, sunk 1942
- SS *Katoomba*, launched on 10 April 1913. Later *Columbia*. Sold to Japanese breakers 22 August 1959
- SS *Alcantara*, launched October 1913, sunk 29 February 1916
- HMHS *Britannic*, improved *Olympic* class, launched on 26 February 1914, sunk by mine 1916
- MS *Akaroa*, maiden voyage 1 July 1914, later *Euripides*, scrapped 1954
- SS *Justicia*, launched on 9 July 1914 as SS *Statendam*, completed April 1917
- RMT *Almanzora*, launched on 19 November 1914, completed September 1915

- SS *Belgenland*, originally SS *Ceric*, launched January 1914, completed 1917 as *Belgic IV*
- HMS *Abercrombie*, launched 15 April 1915, scrapped 1927
- RMS *Regina*, launched 1917, entered passenger service 1919, renamed *Westerland*, scrapped 1947
- SS *Venusia*, in service 1918, sister-ships; first freighters ordered from H+W by Cunard Line
- SS *Varentia*, in service 1918
- RMS *Arundel Castle*, launched 11 September 1919, scrapped 1959
- SS *Doric*, launched 1922, scrapped November 1935
- RMS *Mooltan*, launched on 15 February 1923
- RMS *Maloja*, launched on 19 April 1923
- SS *Minnewaska*, maiden voyage 1 September 1923
- RMMV *Asturias*, launched on 7 July 1925, completed 21 February 1926
- RMMV *Carnarvon Castle*, launched on 14 January 1926, maiden voyage 16 July 1926
- MS *Alcantara*, launched 1927, scrapped 1958
- SS *Laurentic*, launched 16 June 1927, torpedoed and sunk 3 November 1940
- SS *Lochness*, launched 6 June 1929; in service July 1929; scrapped 1973
- MV *Llangibby Castle*, launched 4 July 1929, maiden voyage 5 December 1929
- SS Duke of Lancaster, launched 1956
- MV St Christopher,launched 1981.
- MV St David IV,launched 1981(renamed Stena Caledonia in 1990.)
- RMS *Britannic*, launched 6 August 1929, scrapped 1960
- RMMV *Winchester Castle*, launched 19 November 1929, maiden voyage 24 October 1930
- MS *Achimota*, launched 17 December 1929, delivered 29 November 1932 as TSMV *Wanganella*
- RMMV *Warwick Castle*, launched 29 April 1930, maiden voyage 30 January 1931
- RMS *Georgic*, launched 1931, maiden voyage June 25, 1932, scrapped 1961
- MV *Highland Patriot*, completed 1932, war loss 1 October 1940
- TS *Duchess of Hamilton*, completed 1932, scrapped 1974
- MV *Waipawa*, completed October 1934
- MV *Wairangi*, completed February 1935
- MV *Stirling Castle*, launched 15 July 1935, maiden voyage 7 February 1936
- RMMV *Athlone Castle*, launched 28 November 1935, maiden voyage 22 May 1936
- MS *Dunnottar Castle*, launched 25 January 1936, maiden voyage 10 July 1936, later *Victoria*
- RMMV *Dunvegan Castle*, launched 26 March 1936
- MV *Walmer Castle*, launched 17 September 1936, completed 30 November 1936
- RMMV *Cape Town Castle*, launched 23 September 1937, completed 31 March 1938
- HMS *Belfast*, launched 17 March 1938, decommissioned 24 August 1963, museum ship since 31 October 1971
- RMMV *Durban Castle*, launched on 14 June 1938
- RMS *Andes*, launched on 7 March 1939, completed September 1939
- RMMV *Pretoria Castle*, launched on 12 October 1939 (later HMS *Pretoria Castle* then RMMV *Warwick Castle*)
- HMS *Gentian*, launched 6 August 1940, scrapped 1947
- RFA *Black Ranger*, launched 22 August 1940, scrapped 1979
- HMS *Adamant*, launched 30 November 1940, scrapped 1970
- RFA *Brown Ranger*, launched 12 December 1940, scrapped 1975
- HMS *Anchusa*, launched 15 January 1941, scrapped 1960
- HMS *Bergamot*, launched 15 February 1941, scrapped 1974
- MV *Empire Grace*, launched on 25 August 1941, completed April 1942
- MV *British Bombardier*, built as *Empire Fusilier*, completed as *Empire Bombardier*, scrapped 1959
- HMS *Black Prince*, launched 27 August 1942, scrapped 1962

- MV *Athelqueen*, launched 24 November 1942 as *Empire Benefit*, later *Mariverda*, scrapped 1961
- HMS *Boxer*, launched 12 December 1942, scrapped 1958
- MV *Empire Abercorn*, launched on 30 December 1944
- MV *Durango*, completed in 1944
- MV *Waiwera*, completed in 1944
- HMCS *Bonaventure*, launched 27 February 1945, scrapped 1971
- INS *Vikrant*, completion & modifications only 1957-61, decommissioned 31 January 1997, now museum in Mumbai
- SS *Athenic*, launched on 26 November 1946, maiden voyage 1 August 1947
- RMS *Pretoria Castle*, launched on 19 August 1947 (later SS *S.A. Oranje*)
- RMS *Edinburgh Castle*, launched on 16 October 1947
- RMS *Parthia*, maiden voyage 10 April 1948. First passenger/cargo liner ordered from H+W by Cunard Line
- RMS *Magdalena*, delivered 1949. Ran aground 25 April 1949 on maiden voyage near Rio de Janeiro and wrecked
- MV *Bloemfontein Castle*, launched on 24 August 1949, scrapped 1989
- SS *Rhodesia Castle*, launched on 5 April 1951, completed 6 October 1951
- SS *Kenya Castle*, launched on 21 June 1951, later SS *Amerikanis*, scrapped 2001
- SS *Braemar Castle*, launched on 5 April 1952
- MV *Cedric*, delivered November 1952
- MV *Cretic*, completed in 1953
- SS *Iberia*, launched on 21 January 1954, maiden voyage 28 September 1954
- SS *Loch Gowan*, completed in 1954
- SS *Southern Cross*, launched on 17 August 1954, delivered 23 February 1955, scrapped 2003
- Port Melbourne, completed in 1955
- SS *Reina Del Mar*, launched on 7 June 1955, delivered April 1956, maiden voyage 3 May 1956
- MV " Ulster Star", Launched 26 February 1956, Scrapped 1979
- SS *Loch Loyal*, completed in 1957
- HMS *Blackpool*, launched 14 February 1957, scrapped 1978
- RMS *Pendennis Castle*, launched on 24 December 1957, maiden voyage 1 January 1959, scrapped 1980
- RMMV *Amazon*, launched July 1959, maiden voyage January 1960
- RMMV *Aragon*, launched on 20 October 1959, maiden voyage 29 April 1960
- SS *Canberra*, launched on 16 March 1960, maiden voyage 6 June 1961
- RMMV *Arlanza*, launched on 13 April 1960, maiden voyage 7 October 1960
- RFA *Fort Victoria*, launched on 4 May 1990
- *Knock An*, shuttle tanker, delivered 1996
- *Glas Dowr*, FPSO conversion, delivered 1996
- *Schiehallion*, newbuild FPSO, delivered 1997
- *Bideford Dolphin*, semi-submersible conversion, delivered 1998
- *Borgland Dolphin*, semi-submersible conversion, delivered 1999
- *Glomar CR Luigs*, newbuild dynamically positioned drillship, delivered 2000
- *Glomar Jack Ryan*, newbuild dynamically positioned drillship, delivered 2000
- Aircraft Carriers
  - HMS *Hercules* (R49),
  - HMS *Formidable* (R67),
  - HMS *Eagle* (R05),
  - HMS *Unicorn* (I72),
  - HMS *Campania* (D48),

- HMS *Glory* (R62),
- HMS *Warrior* (R31),
- HMCS *Magnificent* (CVL 21),
- HMCS *Bonaventure* (CVL 22),
- HMS *Centaur* (R06),
- HMS *Bulwark* (R08),
- Cruisers
  - HMS *Penelope* (97),
  - HMS *Belfast* (C35),
  - HMS *Black Prince* (81),
  - HMCS *Ontario* (C53),
- Oil tankers
  - *British Destiny*, British Tanker Company, 1937
  - *British Fidelity*, British Tanker Company, 1938
  - *British Integrity*, British Tanker Company, 1937
  - *British Merit*, British Tanker Company, 1942
  - *British Might*, British Tanker Company, 1945
  - *British Patience*, British Tanker Company, 1943
  - *British Power*, British Tanker Company, 1936
  - *British Security*, British Tanker Company, 1937
  - *British Trust*, British Tanker Company, 1939
  - *British Vigilance*, British Tanker Company, 1942

Earlier ships (number 1 to 150, years 1869 to 1882) are listed by McCluskie (1998:160).[13]

See also Category:Belfast-built ships

References

[1] http://www.harland-wolff.com/

[2] Moss, M. and Hume, J.R. (1986). *Shipbuilders to the World: 125 years of Harland and Wolff, Belfast 1861-1986*. Belfast: Blackstaff Press. pp. xvii, 601 p. ISBN 0-85640-343-1.

[3] "Britain could lead world in offshore wind power" (http://www.telegraph.co.uk/sponsored/business/workingforchange/8323727/Britain-could-lead-world-in-offshore-wind-power.html). *The Telegraph*. February 14, 2011. .

[4] The Guardian - Harland & Wolff locks horns with DTI (http://www.guardian.co.uk/Northern_Ireland/Story/0,2763,190391,00.html)

[5] McCarthy, Michael (2008-01-24). "Britain will need 12,500 wind farms to satisfy EU targets" (http://www.independent.co.uk/environment/climate-change/britain-will-need-12500-wind-farms-to-satisfy-eu-targets-773145.html). The Independent. . Retrieved 2008-10-07.

[6] Harrison, Claire (2008-06-02). "Breath of fresh air for H&W with wind turbine venture" (http://www.belfasttelegraph.co.uk/business/business-news/breath-of-fresh-air-for-hampw-with-wind-turbune-venture-13506748.html). Belfast Telegraph. . Retrieved 2008-10-07.

[7] Henry McDonald, 31 March 2008, The Guardian - Tidal power comes to Northern Ireland (http://www.guardian.co.uk/environment/2008/mar/31/tidal.power)

[8] Turbine contract boost for Harland and Wolff (http://www.insideireland.ie/index.cfm/section/news/ext/harlandandwollf001/category/1067)

[9] "Harland & Wolff Archive, Public Record Office of Northern Ireland" (http://www.rascal.ac.uk/index.php?CollectionID=199&navOp=associated&navVar=88). *RASCAL: Research and Special Collections Available Locally (Ireland)*. . Retrieved 15 December 2011.

[10] http://www.proni.gov.uk/introduction__harland_and_wolff_d2805.pdf

[11] http://www.nmni.com/Home/Shop/Products/Photographs/Harland-and-Wolff-Collection-%28Welch-Collection%29

[12] PRONI (2007). *Introduction Harland and Wolff Papers* (http://www.proni.gov.uk/introduction__harland_and_wolff_d2805.pdf). .

[13] McCluskie, Tom (1998). *Ships from the Archives of Harland & Wolff: Builders of the Titanic*. London: PRC Publishing Ltd. pp. 160. ISBN 1 85648 467.

[14] Passenger lists and Emigrant ships from Norway-Heritage - Shipyard: Harland & Wolff (http://www.norwayheritage.com/p_shipbuild.asp?sb=hwo)

[15] Experience of a UK Shipyard in the 1990s Offshore Market, J. MacGregor, RINA, W272, 2001

[16] http://en.wikipedia.org/wiki/Harland_and_wolff

10. Queen of the Sea, George McWhirter, Oberon Press, 1976,ISBN 0-88750-198-2, Poems about the Belfast Shipyard

External links

- Harland and Wolff Heavy Industries (http://www.harland-wolff.com)
- Harland and Wolff On Titanic-Titanic.com (http://www.titanic-titanic.com/harland_and_wolff.shtml)
- [[Encyclopedia Titanica (http://www.encyclopedia-titanica.org/key_list.php?key=Harland+&+Wolff)] : Harland & Wolff]
- *Nationwide* program feat. a report on Harland and Wolff's bridge-building in the Republic of Ireland (http://www.rte.ie/news/2004/0526/nationwide.html) (Real player required to watch)]
- *Bad Management* program which resulted in a huge crane collapsing. The management currently is still undecided why their crane actually fell the ground. (http://news.bbc.co.uk/1/hi/northern_ireland/6559917.stm)
- Second World War online resource for NI (http://www.secondworldwarni.org)
- H&W Welders FSC (http://www.weldersfc.com)
- PPRNI (http://www.prestigepropertyrepairs.co.uk)

Gross_tonnage

Gross tonnage (often abbreviated as **GT**, **G.T.** or **gt**) is a unitless index related to a ship's overall internal volume. Gross tonnage is different from gross register tonnage.[1] Neither gross tonnage nor gross register tonnage is a measure of the ship's displacement (mass) and should not be confused with terms such as deadweight tonnage or displacement.

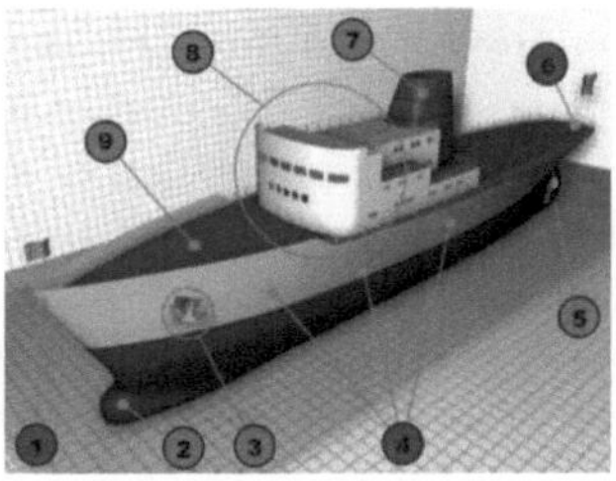

Gross tonnage is calculated by measuring a ship's volume (from keel to funnel, to the outside of the hull framing) and applying a mathematical formula.

Gross tonnage, along with net tonnage, was defined by *The International Convention on Tonnage Measurement of Ships, 1969*, adopted by the International Maritime Organization in 1969, and came into force on July 18, 1982. These two measurements replaced gross register tonnage (GRT) and net register tonnage (NRT). Gross tonnage is calculated based on "the moulded volume of all enclosed spaces of the ship" and is used to determine things such as a ship's manning regulations, safety rules, registration fees and port dues, whereas the older gross register tonnage is a measure of the volume of certain enclosed spaces.

History

The International Convention on Tonnage Measurement of Ships, 1969 was adopted by IMO in 1969.[2] The Convention mandated a transition from the former measurements of gross register tonnage (grt) and net register tonnage (nrt) to gross tonnage (GT) and net tonnage (NT).[2] It was the first successful attempt to introduce a universal tonnage measurement system.[2]

Various methods were previously used to calculate merchant ship tonnage, but they differed significantly and one single international system was needed.[2] Previous methods traced back to George Moorsom of Great Britain's Board of Trade who devised one such method in 1854.[2]

The tonnage determination rules apply to all ships built on or after July 18, 1982.[2] Ships built before that date were given 12 years to migrate from their existing gross register tonnage (GRT) to use of GT and NT.[2] The phase-in period was provided to allow ships time to adjust economically, since tonnage is the basis for satisfying manning regulations and safety rules.[2] Tonnage is also the basis for calculating registration fees and port dues.[2] One of the Convention's goals was to ensure that the new calculated tonnages "did not differ too greatly" from the traditional gross and net register tonnages.[2]

Both GT and NT are obtained by measuring ship's volume and then applying a mathematical formula.[2] Gross tonnage is based on "the moulded volume of all enclosed spaces of the ship" whereas net tonnage is based on "the moulded volume of all cargo spaces of the ship."[2] In addition, a ship's net tonnage is constrained to be no less than 30% of her gross tonnage.[2]

Calculation

The gross tonnage calculation is defined in Regulation 3 of Annex 1 of *The International Convention on Tonnage Measurement of Ships, 1969* [3]. It is based on two variables:

- *V*, the ship's total volume in cubic meters (m^3), and
- *K*, a multiplier based on the ship volume.

The value of the multiplier *K* varies in accordance with a ship's total volume (in cubic metres) and is applied as a kind of reduction factor in determining the gross tonnage value - which does not have a unit such as cubic metres or tons. For smaller ships, *K* is smaller, for larger ships, *K* is larger. *K* ranges from 0.22 to 0.32 and is calculated with a formula which uses the common or base-10 logarithm:

Once *V* and *K* are known, gross tonnage is calculated using the formula, whereby GT is a function of V:

As an example, we can calculate the gross tonnage of a ship with 10,000 m^3 total volume.

Then the gross tonnage is calculated:

See also

- Tonnage
- Moorsom System
- Gross register tonnage
- Compensated gross tonnage
- Ton
- List of world's largest ships by gross tonnage

Notes

[1] The gross register ton (GRT) is a unit of volume defined as 100 cubic feet (about 2.83 m³).

[2] International Maritime Organization, 2002.

[3] http://www.admiraltylawguide.com/conven/tonnage1969.html

References

- "International Convention on Tonnage Measurement of Ships, 1969" (http://www.admiraltylawguide.com/conven/tonnage1969.html). 1969-06-23. Retrieved 2008-03-23.
- "International Convention on Tonnage Measurement of Ships, 1969" (http://www.imo.org/Conventions/contents.asp?topic_id=259&doc_id=685). International Maritime Organization. 2002. Retrieved 2008-03-23.
- Hayler, William B. (2003). *American Merchant Seaman's Manual*. Cornell Maritime Press. ISBN 0-87033-549-9..
- Turpin, Edward A.; McEwen, William A. (1980). *Merchant Marine Officers' Handbook* (4th ed.). Centreville, MD: Cornell Maritime Press. ISBN 978-0-87033-056-8.

Fishguard

Fishguard	
Welsh: *Abergwaun*	

Lower Fishguard

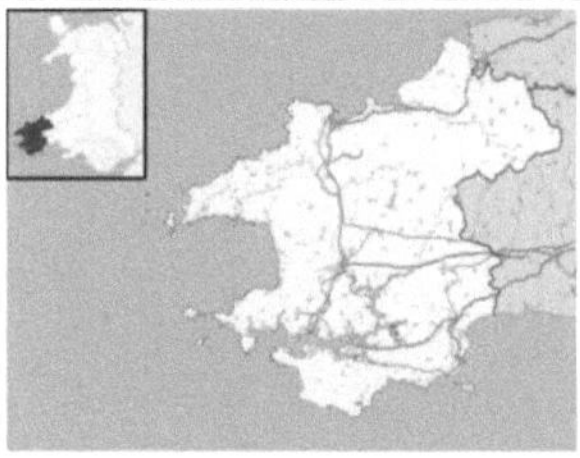

Fishguard

Fishguard shown within Pembrokeshire

Population	**Expression error: "5,043" must be numeric**Template:Infobox UK place/trap[1]
OS grid reference	SM955375
Principal area	Pembrokeshire
Ceremonial county	Dyfed
Country	Wales
Sovereign state	United Kingdom
Post town	FISHGUARD
Postcode district	SA65
Dialling code	01348
Police	Dyfed-Powys
Fire	Mid and West Wales
Ambulance	Welsh
EU Parliament	Wales
UK Parliament	Preseli Pembrokeshire
Welsh Assembly	Preseli Pembrokeshire

Fishguard (Welsh: *Abergwaun*, meaning "Mouth of the River Gwaun") is a coastal town in Pembrokeshire, south-west Wales, with a population of 3,300 (est. 2006). The community of **Fishguard and Goodwick** had a population of 5043 at the 2001 census.

The town of Fishguard (proper) is divided into two parts, the main town of Fishguard and Lower Fishguard. Lower Fishguard (locally known as 'Lower Town') is situated where the River Gwaun meets the sea in a deep valley. It is a typical fishing village with a short tidal quay. The settlement stretches along the north slope of the valley.

The main town contains the parish church, the High Street and most of the modern development, and lies upon the hill to the south of Lower Fishguard, to which it is joined by a steep and winding hill. The western part of the town, facing Goodwick, grew up in the first decade of the 20th century with the development of the harbour.

History

Fishguard is within the historic Welsh cantref of Cemais, and part of the Welsh province of Dyfed, within the historic Principality of Deheubarth.[2] The coasts of Wales were subject to Norse raids during the Viking Era, and in the latter part of the 10th century Norse trading posts and settlements emerged within Dyfed, with Fishguard established sometime between 950 and 1000 AD.[2]

The town was surrounded by secure castle walls and the name *Fishguard* derives from old Norse *fiskigarðr* meaning "fish catching enclosure",[3] indicating that there may have been a Scandinavian trading post, although no evidence has been found.[4] Fishguard was a marcher borough and in 1603 was described as one of five Pembrokeshire boroughs overseen by a portreeve.[5] The Norman settlement lay along what is now High Street between the church at its north end and the remains of a Norman motte at its south end. Lower Fishguard developed as a herring fishery and port, trading with Ireland, Bristol and Liverpool. In the late 18th century it had 50 coasting vessels, and exported oats and salt herring.[6] In 1779, the port was raided by the privateer Black Prince, which bombarded the town when the payment of a £1,000 ransom was refused. As a result, Fishguard Fort was completed in 1781, overlooking Lower Fishguard.[7] The port declined in the latter half of the 19th century.

Fishguard's ancient *Royal Oak* pub was the site of the signing of surrender after the Battle of Fishguard in 1797, said to be the last invasion of Britain, when a force of 1,400 French soldiers landed near Fishguard but surrendered two days later.[8] The story is told in the Fishguard Tapestry created for the 200th anniversary to emulate the Bayeux Tapestry, and displayed in a hall near the town centre. The 19th century vicar of Fishguard, the Rev Samuel Fenton, wrote the book 'The History of Pembrokeshire'. The ancient Parliamentary Borough of Fishguard was contributary to the Borough of Haverfordwest. In 1907, it was created an urban district, which included Goodwick from 1934 until it was abolished in 1974. During the Second World War, the Fishguard Bay Hotel was Station IXc of Special Operations Executive where submersibles were tested in Fishguard Bay.

Since 1995, the town of Loctudy (Breton: *Loktudi*) in Brittany, France has been twinned with Fishguard.

Geography

The town is situated at the back of a north facing bay known as Fishguard Bay (Welsh: Bae Abergwaun) which offers protection from waves generated by prevailing westerly winds. It has a relatively mild climate due to its coastal position. The winds coming from the west or south-west have a determining influence on temperature and precipitation.

Wildlife around Fishguard is rich in flora and fauna: it shows a wide variety of colourful wild flowers and sea mammals including the grey seal, and even porpoises and dolphins. The local birdlife include Eurasian Curlew, Common Redshank and Sanderling regularly foraging in the lower fishguard harbour and European Stonechat, Great Cormorant and Northern Fulmar can be seen from the coastal path.

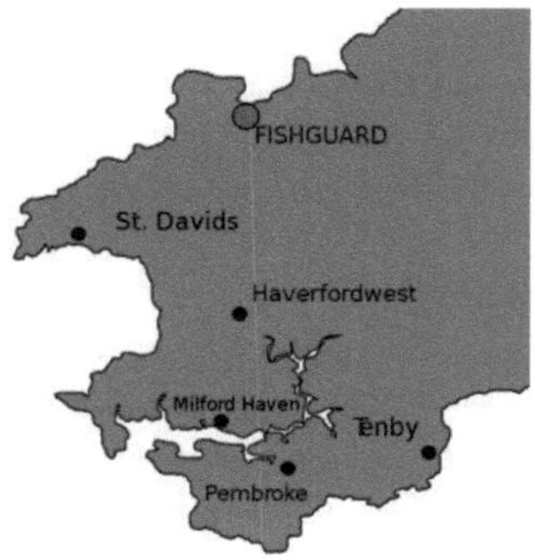

Fishguard in Pembrokeshire

Demography

According to the United Kingdom Census 2001, Fishguard had 3,193 inhabitants and 1,465 households. In 2001, 39.8% of the population could speak Welsh. This compares with 58.9% in 1951 and 90.3% in 1901. The population of 3,193 breaks down as follows:

Age Distribution	Fishguard	Pembrokeshire
0–4 years	5.8%	5.8%
5–15 years	13.0%	14.6%
16–19 years	3.7%	4.8%
20–44 years	24.4%	28.4%
45–64 years	25.2%	27.2%
65+ years	27.9%	19.2%

Landmarks

Outside of Fishguard there is a stone monument commemorating the signing of the Peace Treaty after the last invasion of Britain in 1797. Women dressed in Welsh costume startled the invaders. Also there is the 19th century parish church of St Mary's containing the grave of the heroine Jemima Nicholas. There is also a Bi-Centenary memorial stone monument in West Street, Fishguard to commemorate the Invasion. A tapestry was created in 1997 to commemorate the invasion and can be viewed free of charge in Fishguard's Town Hall [9].

Fishguard has many hotels and is the main shopping town of North Pembrokeshire with a busy Thursday market in the Town Hall.

Fishguard hosted the National Eisteddfod in 1936 and 1986.

Fishguard has a thriving Round Table with 20 members doing all sorts of good work including running the Fishguard & Goodwick Carnival which has been voted the most popular community event.

The Gwaun Lodge of the Royal Antediluvian Order of Buffaloes, is also a thriving charitable organisation within the community who host a number of sponsored events and other community works throughout the year thanks to the influx of younger brethren to the Order.

Fishguard has a 180-seat cinema/theatre called Theatr Gwaun which provides a venue for film, music and live theatre.

Transport

A regular ferry leaves for Rosslare in Ireland from the port of Fishguard Harbour (not actually in Fishguard, but a mile away at Goodwick). Fishguard is the terminus of the A40 London to Fishguard trunk road. It is in the Pembrokeshire Coast National Park and on the Pembrokeshire Coast Path. Fishguard is served by train at Fishguard Harbour railway station.

In the media

Fishguard has acquired a reputation as a result of "Hugh Pugh", a comic character in the Welsh TV series *Barry Welsh is Coming*, who reports from Fishguard and constantly points out the rivalry between Fishguard and Haverfordwest.

Fishguard's *Royal Oak* pub appeared in the film *I'll Sleep When I'm Dead* starring Clive Owen.

Lower Fishguard was used as "Llareggub" in the film of Dylan Thomas's *Under Milk Wood*, starring Richard Burton, Elizabeth Taylor and Peter O'Toole.

The film *Moby Dick* (starring Gregory Peck) was filmed there in 1955.

In sci-fi comic *2000 AD*, Fishguard was the landing point for the Allied liberation of the United Kingdom in *Savage Book 5*.

Notable people

See Category:People from Fishguard

- Mark Delaney, footballer, grew up in Fishguard
- Cerys Matthews, lead singer, Catatonia, went to Fishguard High School and now lives locally
- Jemima Nicholas, single handedly captured 12 French soldiers in 1797, armed only with a pitchfork

See also

- Fishguard Folk Festival

References

[1] Fishguard Ward, 2001 census

[2] Davies, John, *A History of Wales*, Penguin, 1994, *foundations of* pgs 17,19, 43, *Migration of the Desi into Demetia*, page 52 *Demetia* 17, 30, 34, *ruling house of* 52, 72, 85, 87, *and the Vikings* pages 85, *relations with Alfred of Wessex*, page 85, *and the Vikings/Northmen* page 98, *and the Normans* 106, 112, 114

[3] Charles, B. G., *The Placenames of Pembrokeshire*, National Library of Wales, Aberystwyth, 1992, ISBN 0-907158-58-7, p 50

[4] Charles, *ibid*, p xxxvi

[5] Owen, George, *The Description of Penbrokshire by George Owen of Henllys Lord of Kemes*, Henry Owen (Ed), London, 1892

[6] Barrett, J. H., *The Pembrokeshire Coast Path*, HMSO, 1974, ISBN 0-11-700336-0, p 44

[7] Sites and Stones: Fishguard Fort, Pembrokeshire (http://sitesandstones.blogspot.com/2009/04/fishguard-fort-pembrokeshire.html)

[8] Latimer, Jon (12 July 2003). "The Battle of Fishguard: The Last Invasion of Great Britain" (http://www.wargame.ch/wc/nwc/newsletter/21st_edition/Newsletter21/fishguard.html). . Retrieved 7 May 2009.

[9] http://www.visitpembrokeshire.com/news/NewsItem.aspx?ID=41

External links

- historic-uk.com An historic account of Fishguard (http://www.historic-uk.com/HistoryUK/Wales-History/Fishguard.htm)
- Photographs and description from Strolling Guides (http://www.strollingguides.co.uk/books/pembrokeshire/places/fishguard.php)
- Panoramic photograph looking down into Lower Fishguard (http://www.gigapan.org/gigapans/75972/)
- Fishguard Music Festival (http://www.fishguardmusicfestival.co.uk)

Rosslare_Harbour

The village of **Rosslare Harbour** (Irish: *Calafort Ros Láir*, meaning "harbour of the middle peninsula") grew up to serve the needs of the harbour of the same name (now called Rosslare Europort), first developed in 1906 by the Great Western Railway and the Great Southern and Western Railway to accommodate steamferry traffic between Great Britain and Ireland. Rosslare Harbour railway station opened on 30 August 1906.[1]

Certificate of the Fishguard and Rosslare Railways and Harbours Company

Although the harbour itself is located close to the previously existing settlement of Ballygeary, it was named after the village of Rosslare, some 4 km away (8 km by road) along the coast.

The village of Ballygeary was divided into two townslands, one known as tin town and the other as straw town. It is believed this was because of the roofs on the houses.

The village has a number of guesthouses, hotels, a Roman Catholic church, a bank and some shops. Just south of the harbour is a small strand leading to *Greenore Point*, where grey seals are usually to be seen.

See also

- List of towns and villages in Ireland
- List of RNLI stations
- Rosslare Europort railway station

References

[1] "Rosslare Harbour" (http://www.railscot.co.uk/Ireland/Irish_railways.pdf). *Railscot - Irish Railways.* . Retrieved 2007-09-07.

Irish_Sea

Irish Sea		
[[Image:Irish Sea satellite image.jpg		Irish Sea } -]]
[[Image:IrishSeaReliefMap.jpg		Irish Sea - Freight and passenger ports shown as red dots. Freight-only ports as blue dots.]] Freight and passenger ports shown as red dots. Freight-only ports as blue dots.
Location	between Great Britain and Ireland	
Coordinates	53°N 5°W	
Basin countries	United Kingdom; Republic of Ireland	
Islands	Anglesey and Holy Island, Isle of Man, Walney, Lambay, Ireland's Eye	
Settlements	(see below)	

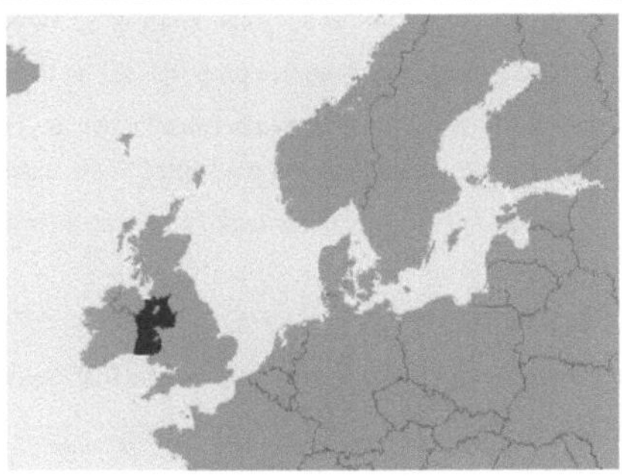
The extent of the Irish Sea

The **Irish Sea** (Irish: *Muir Éireann,*[1] Scottish Gaelic: *Muir Èireann,*[2] Manx: *Y Keayn Yernagh,*[3] Welsh: *Môr Iwerddon*, Scots: *Erse Sea*) separates the islands of Ireland and Great Britain. It is connected to the Celtic Sea in the south by St George's Channel, and to the Atlantic Ocean in the north by the North Channel. Anglesey is the largest island within the Irish Sea, followed by the Isle of Man. The sea is occasionally, but rarely, referred to as the **Manx Sea** (Irish: *Muir Meann,*[4] Scottish Gaelic: *Muir Mhanainn*, Manx: *Mooir Vannin*).[5] [6] [7]

The sea is of significant economic importance to regional trade, shipping and transport, fishing, and power generation in the form of wind power and nuclear plants. Annual traffic between Great Britain and Ireland amounts to over 12 million passengers and 17 million tonnes of traded goods.

Extent

The International Hydrographic Organization defines the limits of the Irish Sea (with St George's Channel) as follows:[8]

> *On the North.* The Southern limit of the Scottish Seas [or Inner Seas off the West Coast of Scotland, defined as "a line joining the South extreme of the Mull of Galloway (54°38'N) in Scotland and Ballyquintin Point (54°20'N) in Ireland"].
>
> *On the South.* A line joining St. David's Head (51°54′N 5°19′W) to Carnsore Point (52°10′N 6°22′W). The sea is bordered by all four constituent countries of the United Kingdom (England, Wales, Scotland and Northern Ireland), along with the Republic of Ireland and the Isle of Man.

Shipping

Unlike Great Britain, Ireland has no tunnel or bridge connection to mainland Europe. Thus the vast majority of heavy goods trade is done by sea. Northern Ireland ports handle 10 million tonnes of goods trade with Great Britain annually, while ports in the Republic of Ireland handle 7.6 million tonnes, representing 50% and 40% respectively of total trade by weight.

The Port of Liverpool handles 32 million tonnes of cargo and 734 thousand passengers a year.[9] Holyhead port handles most of the passenger traffic from Dublin and Dún Laoghaire ports, as well as 3.3 million tonnes of freight.[10]

Ports in the Republic handle 3,600,000 travellers crossing the sea each year, amounting to 92% of all Irish Sea travel.[11] This has been steadily dropping for a number of years (20% since 1999), probably as a result of low cost airlines.

Ferry connections from Great Britain to Ireland across the Irish Sea include the routes from Swansea to Cork, Fishguard and Pembroke to Rosslare, Holyhead to Dún Laoghaire, Holyhead to Dublin, Stranraer to Belfast and Larne, and Cairnryan to Larne. There is also a connection between Liverpool and Belfast via the Isle of Man or direct from Birkenhead. The world's largest car ferry, *Ulysses*, is operated by Irish Ferries on the Dublin Port–Holyhead route; Stena Line also operates between Britain and Ireland. The Port of Barrow-in-Furness, despite being one of Britain's largest shipbuilding centres and being home to the United Kingdom's only submarine-building complex, is only a minor port.

"Irish Sea" is also the name of one of the BBC's Shipping Forecast areas defined by the coordinates: 54°50′N 05°05′W 54°45′N 05°45′W 52°30′N 06°15′W 52°00′N 05°05′W

There have been various tentative proposals for an Irish Sea Tunnel.

During World War I the Irish Sea became known as "U-boat Alley", because the U-boats moved their emphasis from the Atlantic to the Irish Sea after the United States entered the war in 1917.[12] [13]

See also: Transport in Ireland, Transport in the United Kingdom, Transport on the Isle of Man

Cities and towns

Below is a list of cities and towns around the Irish Sea coasts in order of size:

Rank	City/Town	County	Region/Province	Population	Country
1	Dublin	County Dublin	Leinster	505,739	Republic of Ireland
2	Liverpool	Merseyside	North West	447,500	England
3	Belfast	County Antrim	Ulster	276,459	Northern Ireland
5	Blackpool	Lancashire	North West	142,900	England
6	Southport	Merseyside	North West	99,456	England
7	Birkenhead	Merseyside	North West	83,729	England
8	Bangor	County Down	Ulster	76,851	Northern Ireland
9	Barrow-in-Furness	Cumbria	North West	71,980	England
10	Wallasey	Merseyside	North West	58,710	England
11	Morecambe	Lancashire	North West	45,000	England
12	Lytham St Annes	Lancashire	North West	41,330	England
13	Drogheda	County Louth	Leinster	35,190	Republic of Ireland
14	Dundalk	County Louth	Leinster	35,085	Republic of Ireland

15	Rhyl	Denbighshire	Clwyd	35,000	Wales
17	Bray	County Wicklow	Leinster	31,901	Republic of Ireland
18	Thornton-Cleveleys	Lancashire	North West	31,157	England
19	Colwyn Bay	Conwy	Clwyd	30,265	Wales
20	Carrickfergus	County Antrim	Ulster	27,201	Northern Ireland
21	Fleetwood	Lancashire	North West	26,840	England
22	Douglas	N/A	Isle of Man	26,218	Isle of Man
20	Workington	Cumbria	North West	25,978	England
23	Whitehaven	Cumbria	North West	25,500	England
24	Dún Laoghaire	Dún Laoghaire–Rathdown	Leinster	23,857	Republic of Ireland
25	Llandudno	Conwy	Clwyd	20,090	Wales
26	Larne	County Antrim	Ulster	18,228	Northern Ireland
27	Wexford	County Wexford	Leinster	18,163	Republic of Ireland

Origin

The Irish Sea has undergone a series of dramatic changes over the last 20,000 years as the last glacial period ended and was replaced by warmer conditions. At the height of the glaciation the central part of the modern sea was probably a long freshwater lake. As the ice retreated 10,000 years ago the lake reconnected to the sea, becoming brackish and then fully saline once again.

Environment

Biodiversity

Brittas Bay, on the County Wicklow coast

The most accessible and possibly the greatest wildlife resource of the Irish Sea lies in its estuaries: particularly the Dee Estuary, the Mersey Estuary, the Ribble Estuary, Morecambe Bay, the Solway Firth, Loch Ryan, the Firth of Clyde, Belfast Lough, Strangford Lough, Carlingford Lough, Dundalk Bay, Dublin Bay and Wexford Harbour. However, a lot of wildlife also depends on the cliffs, salt marshes and sand dunes of the adjoining shores, the seabed and the open sea itself.

The information on the invertebrates of the seabed of the Irish Sea is rather patchy because it is difficult to survey such a large area, where underwater visibility is often poor and information often depends upon looking at material brought up from the seabed in mechanical grabs. However, the groupings of animals present depend to a large extent on whether the seabed is composed of rock, boulders, gravel, sand, mud or even peat. In the soft sediments seven types of community have been provisionally identified, variously dominated by brittle-stars, sea urchins, worms, mussels, tellins, furrow-shells, and tower-shells.

Parts of the bed of the Irish Sea are very rich in wildlife. The seabed southwest of the Isle of Man is particularly noted for its rarities and diversity,[14] as are the horse mussel beds of Strangford Lough. Scallops and queen scallops are found in more gravelly areas. In the estuaries, where the bed is more sandy or muddy, the number of species is smaller but the size of their populations is larger. Brown shrimp, cockles and edible mussels support local fisheries in Morecambe Bay and the Dee Estuary and the estuaries are also important as nurseries for flatfish, herring and sea bass. Muddy seabeds in deeper waters are home to populations of the Dublin Bay prawn, also known as "scampi".[15]

The open sea is a complex habitat in its own right. It exists in three spatial dimensions and also varies over time and tide. For example, where freshwater flows into the Irish Sea in river estuaries its influence can extend far offshore as the freshwater is lighter and "floats" on top of the much larger body of saltwater until wind and temperature changes mix it in. Similarly, warmer water is less dense and seawater warmed in the inter-tidal zone may "float" on the colder offshore water. The amount of light penetrating the seawater also varies with depth and turbidity. This leads to differing populations of plankton in different parts of the sea and varying communities of animals that feed on these populations. However, increasing seasonal storminess leads to greater mixing of water and tends to break down these divisions, which are more apparent when the weather is calm for long periods.

Plankton includes bacteria, plants (phytoplankton) and animals (zooplankton) that drift in the sea. Most are microscopic, but some, such as the various species of jellyfish and sea gooseberry, can be much bigger.

Diatoms and dinoflagellates dominate the phytoplankton. Although they are microscopic plants, diatoms have hard shells and dinoflagellates have little tails that propel them through the water. Phytoplankton populations in the Irish Sea have a spring "bloom" every April and May, when the seawater is generally at its greenest.

Crustaceans, especially copepods, dominate the zooplankton. However, many animals of the seabed, the open sea and the seashore spend their juvenile stages as part of the zooplankton. The whole plankton "soup" is vitally important, directly or indirectly, as a food source for most species in the Irish Sea, even the largest. The enormous basking shark, for example, lives entirely on plankton and the leatherback turtle's main food is jellyfish.

A colossal diversity of invertebrate species live in the Irish Sea and its surrounding coastline, ranging from flower-like fan-worms to predatory swimming crabs to large chameleon-like cuttlefish.[15] Some of the most significant for other wildlife are the reef-building species like the inshore horse mussel of Strangford Lough and the inter-tidal honeycomb worm of Morecambe Bay, Cumbria and Lancashire. These build up large structures over many years and, in turn, provide surfaces, nooks and crannies where other marine animals and plants may become established and live out some or all of their lives.

There are quite regular records of live and stranded leatherback turtle in and around the Irish Sea. This species travels north to the waters off the British Isles every year following the swarms of jellyfish that form its prey. Loggerhead turtle, Ridley sea turtle and green turtle are found very occasionally in the Irish Sea but are generally unwell or dead when discovered. They have strayed or been swept out of their natural range further south into colder waters.[16]

The estuaries of the Irish Sea are of international importance for birds. They are vital feeding grounds on migration flyways for shorebirds travelling between the Arctic and Africa. Others depend on the milder climate as a refuge when continental Europe is in the grip of winter.[15]

Twenty-one species of seabird are reported as regularly nesting on beaches or cliffs around the Irish Sea. Huge populations of the sea duck, common scoter, spend winters feeding in shallow waters off eastern Ireland, Lancashire and North Wales.[15]

Whales, dolphins and porpoises all frequent the Irish Sea, but knowledge of how many there may be and where they go is somewhat sketchy. About a dozen species have been recorded since 1980, but only three are seen fairly often. These are the harbour porpoise, bottlenose dolphin and common dolphin. The more rarely seen species are minke whale, fin whale, sei whale, sperm whale, northern bottlenose whale, long-finned pilot whale, orca, white-beaked dolphin, striped dolphin and Risso's dolphin.[15]

The common or harbour seal and the grey seal are both resident in the Irish Sea. Common seals breed in Strangford Lough, grey seals in southwest Wales and, in small numbers, on the Isle of Man. Grey seals haul out, but do not breed, off Hilbre and Walney islands, Merseyside, the Wirral, Barrow-in-Furness Borough, and Cumbria.[15]

Radioactivity

The Irish Sea has been described by Greenpeace as the most radioactively contaminated sea in the world with some "eight million litres of nuclear waste" discharged into it each day from Sellafield reprocessing plants, contaminating seawater, sediments and marine life.[17]

Low-level radioactive waste has been discharged into the Irish Sea as part of operations at Sellafield since 1952. The rate of discharge began to accelerate in the mid- to late 1960s, reaching a peak in the 1970s and generally declining significantly since then. As an example of this profile, discharges of plutonium (specifically ^{241}Pu) peaked in 1973 at 2,755TBq[18] falling to 8.1 TBq by 2004.[19] Improvements in the treatment of waste in 1985 and 1994 resulted in further reductions in radioactive waste discharge although the subsequent processing of a backlog resulted in increased discharges of certain types of radioactive waste. Discharges of technetium in particular rose from 6.1 TBq in 1993 to a peak of 192TBq in 1995 before dropping back to 14TBq in 2004.[18] [19] In total 22PBq of ^{241}Pu was discharged over the period 1952 to 1998.[20] Current rates of discharge for many radionuclides are at least 100 times lower than they were in the 1970s.[21]

Dublin Bay on the west coast of Irish Sea

Analysis[22] [23] of the distribution of radioactive contamination after discharge reveals that mean sea currents result in much of the more soluble elements such as caesium being flushed out of the Irish Sea through the North Channel about a year after discharge. Measurements of technetium concentrations post-1994 has produced estimated transit times to the North Channel of around six months with peak concentrations off the northeast Irish coast occurring 18–24 months after peak discharge. Less soluble elements such as plutonium are subject to much slower redistribution. Whilst concentrations have declined in line with the reduction in discharges they are markedly higher in the eastern Irish Sea compared to the western areas. The dispersal of these elements is closely associated with sediment activity, with muddy deposits on the seabed acting as sinks, soaking up an estimated 200 kg of plutonium.[24] The highest concentration is found in the eastern Irish Sea in sediment banks lying parallel to the Cumbrian coast. This area acts as a significant source of wider contamination as radionuclides are dissolved once again. Studies have revealed that 80% of current sea water contamination by caesium is sourced from sediment banks, whilst plutonium levels in the western sediment banks between the Isle of Man and the Irish coast are being maintained by contamination redistributed from the eastern sediment banks.

The consumption of seafood harvested from the Irish Sea is the main pathway for exposure of humans to radioactivity.[25] The environmental monitoring report for the period 2003 to 2005 published by the Radiological Protection Institute of Ireland (RPII) reported that in 2005 average quantities of radioactive contamination found in seafood ranged from less than 1Bq/kg for fish to under 44Bq/kg for mussels.[26] Doses of man-made radioactivity received by the heaviest consumers of seafood in Ireland in 2005 was 1.10μSv.[27] This compares with a corresponding dosage of radioactivity naturally occurring in the seafood consumed by this group of 148μSv and a total average dosage in Ireland from all sources of 3620μSv.[28] In terms of risk to this group, heavy consumption of seafood generates a 1 in 18 million chance of causing cancer. The general risk of contracting cancer in Ireland is 1 in 522. In the UK, the heaviest seafood consumers in Cumbria received a radioactive dosage attributable to Sellafield discharges of 0.22mSv (220μSv) in 2005.[29] This compares to average annual dose of naturally sourced radiation received in the UK of 2.23mSv (2230μSv).[30]

Also see Beaufort's Dyke.

Oil and gas exploration

East Irish Sea Basin

With 7.5 trillion cubic feet (210 km³) of natural gas and 176 million barrels (28,000,000 m³) of petroleum estimated by the field operators as initially recoverable hydrocarbon reserves from eight producing fields (DTI, 2001), the East Irish Sea Basin is at a mature exploration phase. Early Namurian basinal mudstones are the source rocks for these hydrocarbons. Production from all fields is from fault-bounded traps of the Lower Triassic formation, principally the aeolian Sherwood Sandstone reservoir, top-sealed by younger Triassic continental mudstones and evaporites. Future mineral exploration will initially concentrate on extending this play, but there remains largely untested potential also for gas and oil within widespread Carboniferous fluvial sandstone reservoirs. This play requires intraformational mudstone seal units to be present, as there is no top-seal for reservoirs subcropping the regional base Permian unconformity in the east of the basin, and Carboniferous strata crop out at the sea bed in the west.

Caernarfon Bay Basin

The Caernarfon Bay basin contains up to 7 km³ of Permian and Triassic syn-rift sediments in an asymmetrical graben that is bounded to the north and south by Lower Paleozoic massifs. Only two exploration wells have been drilled so far, and there remain numerous undrilled targets in tilted fault block plays. As in the East Irish Sea Basin, the principal target reservoir is the Lower Triassic, Sherwood Sandstone, top-sealed by younger Triassic mudstones and evaporites. Wells in the Irish Sector to the west have demonstrated that pre-rift, Westphalian coal measures are excellent hydrocarbon source rocks, and are at peak maturity for gas generation (Maddox et al., 1995). Seismic profiles clearly image these strata continuing beneath a basal Permian unconformity into at least the western part of the Caernarfon Bay Basin. The timing of gas generation presents the greatest exploration risk. Maximum burial of, and primary gas migration from, the source rocks could have terminated as early as the Jurassic, whereas many of the tilted fault blocks were reactivated or created during Paleogene inversion of the basin. However, it is also possible that a secondary gas charge occurred during regional heating associated with intrusion of Paleogene dykes, such as those that crop out nearby on the coastline of north Wales. (Floodpage et al., 1999) have invoked this second phase of Paleogene hydrocarbon generation as an important factor in the charging of the East Irish Sea Basin's oil and gas fields. It is not clear as yet whether aeromagnetic anomalies in the southeast of Caernarfon Bay are imaging a continuation of the dyke swarm into this area too, or whether they are instead associated with deeply buried Permian syn-rift volcanics. Alternatively, the fault block traps could have been recharged by exsolution of methane from formation brines as a direct result of the Tertiary uplift (cf. Doré and Jensen, 1996).

The Cardigan Bay Basin

The Cardigan Bay Basin forms a continuation into British waters of Ireland's North Celtic Sea Basin, which has two producing gas fields. The basin comprises a south-easterly deepening half-graben near the Welsh coastline, although its internal structure becomes increasingly complex towards the southwest. Permian to Triassic, syn-rift sediments within the basin are less than 3 km thick and are overlain by up to 4 km of Jurassic strata, and locally also by up to 2 km of Paleogene fluvio-deltaic sediments. The basin has a proven petroleum system, with potentially producible gas reserves at the Dragon discovery near the UK/ROI median line, and oil shows in a further three wells. The Cardigan Bay Basin contains multiple reservoir targets, which include the Lower Triassic (Sherwood Sandstone), Middle Jurassic shallow marine sandstones and limestone (Great Oolite), and Upper Jurassic fluvial sandstone, the reservoir for the Dragon discovery. The most likely hydrocarbon source rocks are Early Jurassic marine mudstones. These are fully mature for oil generation in the west of the British sector, and are mature for gas generation nearby in the Irish sector. Gas-prone, Westphalian pre-rift coal measures may also be present at depth locally. The Cardigan Bay Basin was subjected to two Tertiary phases of compressive uplift, whereas maximum burial that terminated primary hydrocarbon generation was probably around the end of the Cretaceous, or earlier if Cretaceous strata, now missing, were never deposited in the basin. Despite the Tertiary structuration, the Dragon discovery has proved that potentially commercial volumes of hydrocarbons were retained at least locally in Cardigan Bay. In addition to undrilled structural traps, the basin contains untested potential for stratigraphic entrapment of hydrocarbons near synsedimentary faults, especially in the Middle Jurassic section.[31] [32]

Liverpool Bay

The Liverpool Bay Development is BHP Billiton Petroleum's largest operated asset. It comprises the integrated development of five offshore oil and gas fields in the Irish Sea:

- Douglas oil field
- Hamilton gas field
- Hamilton North gas field
- Hamilton East gas field
- Lennox oil and gas field

Oil is produced from the Lennox and Douglas fields. It is then treated at the Douglas Complex and piped 17 kilometres to an oil storage barge ready for export by tankers. Gas is produced from the Hamilton, Hamilton North and Hamilton East reservoirs. After initial processing at the Douglas Complex the gas is piped by subsea pipeline to the Point of Ayr gas terminal for further processing. The gas is then sent by onshore pipeline to PowerGen's combined cycle gas turbine power station at Connah's Quay. PowerGen is the sole purchaser of gas from the Liverpool Bay development.

The Liverpool Bay development comprises four offshore platforms. Offshore storage and loading facilities. The onshore gas processing terminal at Point of Ayr. Production first started at each filed as follows: Hamilton North in 1995, Hamilton in 1996, Douglas in 1996, Lennox (oil only) in 1996 and Hamilton East 2001. The first contract gas sales were in 1996.

The quality of the water in Liverpool Bay was historically contaminated by dumoing of sewage sludge at sea[33] but this practice became illegal in December 1988 and no further sludge was deposited after that date[34]

Dalkey Island Exploration Prospect

Previous exploration drilling in the Kish Bank Basin has confirmed the potential for petroleum generation with oil shows seen in a number of wells together with natural hydrocarbon seeps recorded from airborne surveys. New analysis of vintage 2-D seismic data has revealed the presence of a large undrilled structural closure at Lower Triassic level situated c. 10 km offshore Dublin. This feature, known as the Dalkey Island exploration prospect, may be prospective for oil, as there are prolific oil productive Lower Triassic reservoirs nearby in the eastern Irish Sea offshore Liverpool. Whilst the Dalkey Island exploration prospect could contain c. 870 MMBO in place, this undrilled prospect still has significant risk and the partners are currently advancing a focused work programme in order to better understand and hopefully mitigate these risks. However, given its location in shallow water and close proximity to shore, the prospect is of great interest as exploration drilling, together with any future development costs, are likely to be low.

Proposed tunnel projects

Discussions of linking Britain to Ireland began in 1895,[35] with an application for £15,000 towards the cost of carrying out borings and soundings in the North Channel to see if a tunnel between Ireland and Scotland was viable. Sixty years later Harford Montgomery Hyde, Unionist MP for North Belfast, called for the building of such a tunnel.[36] A tunnel project has been discussed several times in the Irish parliament.[37] [38] [39] [40] The idea for such a 34 km (21 mi) long rail bridge or tunnel, continues to be mooted.[41]

Several potential Irish Sea tunnel projects have been proposed, most recently the "Tusker Tunnel" between the ports of Rosslare and Fishguard proposed by The Institute of Engineers of Ireland in 2004.[42] [43] A different proposed route between Dublin and Holyhead was proposed in 1997 by the British engineering firm Symonds. Either tunnel, at 80 kilometres (**unknown operator: u'strong'** mi), would be by far the longest in the world, and would cost an estimated €20 billion.

Wind power

An offshore wind farm was developed on the Arklow Bank,[44] Arklow Bank Wind Park, about 10 km off the coast of County Wicklow in the south Irish Sea. The site currently has seven GE 3.6 MW turbines, each with 104 m diameter rotors, the world's first commercial application of offshore wind turbines over three megawatts in size. The operating company, Airtricity, has indefinite plans for nearly 100 further turbines on the site.

Barrow Offshore windfarm, off Walney Island

Further wind turbine sites include:

- The North Hoyle site 5 miles (**unknown operator: u'strong'** km) off the coast from Rhyl and Prestatyn in North Wales, containing thirty 2 MW turbines.[45] operated by NPower Renewables
- Burbo Bank site 10 km off the north Wirral coast
- Robin Rigg Wind Farm in the Solway Firth
- Thirty 90 m 3 MW turbines are operating in a wind farm 7 km off the coast of Walney Island.[46]
- Turbines are being erected off the coast of Clogherhead (to be called the Oriel Wind Farm)[47]

References

[1] Focal.ie (http://www.focal.ie/Search.aspx?term=Muir_Éireann&lang=2)

[2] Cambridge Medieval Celtic Studies , Issues 33–35 University of Cambridge (Gran Bretaña). Department of Anglo-Saxon, Norse and Celtic 1997

[3] "Ellan Vannin" (http://www.liv.ac.uk/manxstudies/laare-studeyryes_manninagh.htm) (in Manx). Centre for Manx Studies ("Laare-Studeyrys Manninagh"). . Retrieved 8 July 2011.

[4] Electronic Dictionary of the Irish Language (http://www.dil.ie/results-list.asp?Fuzzy=0&cv=1&searchtext=(id contains M) and (column contains 103)&sortField=ID&sortDIR=65602&respage=0&resperpage=10&bhcp=1)

[5] Bannerman, David Armitage (1963). *The Birds of the British Isles: Volume 12*. Edinburgh: Oliver and Boyd. pp. 84.

[6] *The Caledonian: Volume 4* (New York: Caledonian Publishing Co.): pp. 25. 1903.

[7] "Irish Sea Facts" (http://www.irishseaconservation.org.uk/facts). Irish Sea Conservation. . Retrieved 3 July 2011.

[8] "Limits of Oceans and Seas, 3rd edition + corrections" (http://www.iho-ohi.net/iho_pubs/standard/S-23/S23_1953.pdf). International Hydrographic Organization. 1971. p. 42 [corrections to page 12]. . Retrieved 6 February 2010.

[9] Port Statistics, (Link) (http://www.merseydocks.co.uk/index2.htm), Mersey Docks Website

[10] UK Port Traffic Highlights: 2002, (pdf) (http://www.maritimestatistics.org/pdfs/port traffic highlights 2002.pdf), UK Maritime Statistics, Dept of Transport

[11] Direct Passenger Movement by Sea from and to Ireland (Republic), (link) (http://www.cso.ie/statistics/passengermovementbysea.htm), Central Statistics Office of Ireland

[12] *U-Boat Alley* by Roy Stokes, published by Compuwreck, ISBN 0-9549186-0-6

[13] The War in Maps: The Irish Sea, (Link) (http://uboat.net/maps/irish_sea.htm), UBoat.net

[14] Barne, J.H., Robson, C.F., Kaznowska, S.S., Doody, J.P., & Davidson, N.C., eds. 1996. Coasts and seas of the United Kingdom. Region 13 Northern Irish Sea: Colwyn Bay to Stranraer, including the Isle of Man. Peterborough, Joint Nature Conservation Committee, ISBN 1-873701-87-X

[15] Irish Sea Study Group Report, Part 1, NATURE CONSERVATION, Liverpool University Press, 1990 ISBN 0-85323-227-X

[16] Irish Sea Leatherback Turtle Project – aiming to understand the populations, origins and behaviour of leatherback turtles in the Irish Sea (http://www.turtle.ie/index.htm) http://www.jellyfish.ie/turtle.asp

[17] Sellafield nuclear reprocessing facility, (Link) (http://www.greenpeace.org.uk/nuclear/sellafield-nuclear-reprocessing-facility), Greenpeace

[18] The Past, Current and Future Radiological Impact of the Sellafield Marine Discharges on the People Living in the Coastal Communities Surrounding the Irish Sea, (Link) (http://publications.environment-agency.gov.uk/pdf/STRP290-e-p.pdf), Environment Agency – Table 3

[19] Monitoring our Environment – Discharges and Monitoring in the UK – Annual Report 2004, (Link) (http://www.britishnucleargroup.com/pdf/BNGSL 2004 Monitoring report final.pdf), British Nuclear Group – Table 2

[20] León Vintró *et al.* (2000), p. 2.

[21] Quality Status Report – Regional QSR III, (Link) (http://www.ospar.org/eng/doc/pdfs/R3C4.pdf), OSPAR – Chapter 4 Chemistry, p64

[22] León Vintró *et al.* (2000), sections 3–4.

[23] McMahon et al., 2005, (Link) (http://www.edpsciences.org/articles/radiopro/pdf/2005/02/p424.pdf?access=ok), Transfer of conservative and non-conservative radionuclides from the Sellafield Nuclear Fuel Reprocessing plant to the coastal waters of Ireland

[24] Quality Status Report – Regional QSR III, (Link) (http://www.ospar.org/eng/doc/pdfs/R3C4.pdf), OSPAR – Chapter 4 Chemistry, p66

[25] Ryan *et al.* (2005), p. 7.

[26] Ryan *et al.* (2005), Table 45.

[27] Ryan *et al.* (2005), p. 26.

[28] Ryan *et al.* (2005), p. 27.

[29] Radioactivity in Food and the Environment 2005, (Link) (http://www.cefas.co.uk/Publications/rife/rife11.pdf), Cefas – p11

[30] Watson et al., 2005 (Link) (http://www.hpa.org.uk/radiation/publications/hpa_rpd_reports/2005/hpa_rpd_001.pdf), Health Protection Agency – Ionising Radiation Exposure of the UK Population: 2005 Review

[31] "Petroleum prospectivity of the principal sedimentary basins on the United Kingdom Continental Shelf" (pdf) (http://www.og.dti.gov.uk/UKpromote/summary_UKCS_2003.pdf), Department of Trade and Industry, 2003

[32] Liverpool Bay, England (Link) (http://www.visitorcentre-bhpbilliton.com/), BHP Oil Ltd

[33] C.Michael Hogan. 2011. *Irish Sea.* eds. P.Saundry & C.Cleveland. Encyclopedia of Earth. National Council for Science and the Environment. Washington DC (http://www.eoearth.org/article/Irish_Sea?topic=49523)

[34] Hansard - Pollution of Liverpool Bay (http://www.publications.parliament.uk/pa/cm200607/cmhansrd/cm070709/text/70709w0003.htm)

[35] "TUNNEL UNDER THE SEA", The Washington Post, 2 May 1897 (Archive link) (http://pqasb.pqarchiver.com/washingtonpost_historical/access/282943272.html?dids=282943272:282943272&FMT=ABS&FMTS=ABS:FT&date=MAY+02,+1897&author=&pub=The+Washington+Post&desc=TUNNEL+UNDER+THE+SEA&pqatl=google)

[36] "An Irishman's Diary" by Wesley Boyd, (Link) (http://www.irishtimes.com/newspaper/opinion/2004/0207/3742943993DIFEB7.html), The Irish Times, Feb 2004 (subscription required)

[37] Written Answers. – Sea Transport, (Link) (http://historical-debates.oireachtas.ie/D/0384/D.0384.198811160027.html), Dáil Éireann – Volume 384 – 16 November 1988

[38] Written Answers. – Irish Sea Railway Ferry, (Link) (http://historical-debates.oireachtas.ie/D/0434/D.0434.199310190089.html), Dáil Éireann – Volume 434 – 19 October 1993

[39] Written Answers. – Ireland-UK Tunnel, (Link) (http://www.oireachtas-debates.gov.ie/plweb-cgi/fastweb?state_id=1140781637&view=oho-view&docrank=44&numhitsfound=315&query=Irish sea tunnel&query_rule=(($query1)<=DATE<=($query2)) AND (($query4)):SPEAKER AND (($query5)):heading AND (($query6)):CATEGORY AND (($query3)):house AND (($query7)):volume AND (($query8)):colnumber AND (($query))&docid=383293&docdb=Debates&dbname=Debates&starthit=25&sorting=none&operator=and&TemplateName=predoc.tmpl&setCookie=1), Dáil Éireann – Volume 517 – 29 March 2000

[40] Written Answers – Transport Projects, (Link) (http://www.oireachtas-debates.gov.ie/plweb-cgi/fastweb?state_id=1140781475&view=oho-view&docrank=4&numhitsfound=315&query=Irish sea tunnel&query_rule=(($query1)<=DATE<=($query2)) AND (($query4)):SPEAKER AND (($query5)):heading AND (($query6)):CATEGORY AND (($query3)):house AND (($query7)):volume AND (($query8)):colnumber AND (($query))&docid=477035&docdb=Debates&dbname=Debates&sorting=none&operator=and&TemplateName=predoc.tmpl&setCookie=1), Dáil Éireann – Volume 597 – 15 February 2005

[41] Bridge to Northern Ireland mooted (http://news.bbc.co.uk/2/hi/uk_news/scotland/south_of_scotland/6958191.stm), BBC News Scotland, 22 August 2007

[42] A Vision of Transport in Ireland in 2050, IEI report (pdf) (http://www.iei.ie/Publications/GetPublicationDetails.pasp?PublicationID=69&Module=Papers&txt_freetext=&RecordsPerPage=1000&PageNumber=1&MenuID=24), The Irish Academy of Engineers, 21 December 2004

[43] "Tunnel 'vision' under Irish Sea" (http://news.bbc.co.uk/1/hi/wales/south_west/4121001.stm). BBC News. 23 December 2004. .

[44] "Arklow Bank Wind Park" (http://www.airtricity.com/ireland/wind_farms/offshore/operating/arklow_bank/index.xml). Airtricity. .

[45] Northhoyle (http://www.natwindpower.co.uk/northhoyle/)

[46] "Barrow Offshore windfarm" (http://www.bowind.co.uk). .

[47] Oriel Wind project status (http://www.orielwind.com/project_status.html)

Bibliography

- Luis León Vintró, Kilian J. Smith, Julie A. Lucey & Peter I. Mitchell (4–6 December 2000). "The environmental impact of the Sellafield discharges" (http://homepage.eircom.net/~radphys/scope.pdf) (PDF). SCOPE-RADSITE Workshop. Brussels.
- R. W. Ryan, A. Dowdall, M. F. Fegan, E. Hayden, K. Kelleher, S. Long, I. McEvoy, L. McKittrick, C. A. McMahon, M. Murray, K. Smith, S. Sequeira, J. Wong & D. Pollard (PDF). *Radioactive Monitoring of the Irish Environment 2003–2005* (http://web.archive.org/web/20090327121438/http://www.rpii.ie/Download/Env_Mon_Rep_03_05.pdf). Radiological Protection Institute of Ireland. Archived from the original (http://www.rpii.ie/Download/Env_Mon_Rep_03_05.pdf) on 27 March 2009.

External links

- Fylde Coast Marine Life Project (http://www.fyldecoastmarinelife.btik.com)
- Irish Sea Conservation Zones (http://www.irishseaconservation.org.uk)
- The Wildlife Trusts Living Seas – Irish Sea (http://www.irishsea.org)

Article Sources and Contributors

' *Source*: http://en.wikipedia.org/w/index.php?title=MS_Stena_Cambria *Contributors*: Dl2000, JaGa, JonEastham, Mandarax, Techspert, William Avery, Woohookitty

Sealink *Source*: http://en.wikipedia.org/w/index.php?title=Sealink *Contributors*: Aquilina, ArtVandelay13, Brutaldeluxe, Btline, Cambria49, Choalbaton, Chris Henniker, Classical geographer, Cliffb, Cod, Dr Greg, Elsecar, Emeraude, Garethhamilton, Gidonb, Gwernol, Hugo999, Ian Dunster, JonEastham, Lee McLoughlin Leicester, Lockley, Matthewross, Mauls, Mddkpp, Mike.cockroft, MrWeeble, Naupigos, Olddemdike, Oranjblud, Penrithguy, Philphos, R'n'B, ScottDavis, Sealinktg, Sionnach1, Sselbor, Suction Man, Topbanana, Trident13, Twydallaer, Vreemdst, WOSlinker, 34 anonymous edits

Ventouris_Ferries *Source*: http://en.wikipedia.org/w/index.php?title=Ventouris_Ferries *Contributors*: AvicAWB, Jeanhousen, Philphos, Stormbay

English_Channel *Source*: http://en.wikipedia.org/w/index.php?title=English_Channel *Contributors*: 193.133.134.xxx, A. di M., A.bracegirdle, AHert, AMX, Aasmunds, Abhin 555, Academic Challenger, Aeusoes1, Afil, Ahoerstemeier, AimalCool, Albert Krantz, Alex43223, AlexiusHoratius, AlfreD123, Alzarian16, Anclation, Andre Engels, Andrew Dalby, Andrewfinn2009, Andrewrp, AndyZ, Animum, AnnaFrance, AnonMoos, Anotheruserhere, Anthony Appleyard, Antriver, Anywikiuser, Aratuk, Arg, Arthena, Asarlaí, AtikuX, Aubadaurada, Aubrey rb, Avenue, AxSkov, AxelBoldt, AzaToth, Backspace, BanyanTree, Bardsandwarriors, Basalisk, Bazonka, Bearsfan85, Bender235, Big floppers, BillMasen, Billydeeuk, Binary TSO, Bjmullan, Bob6, Bobblewik, Bobbo, Bobo192, Bodnotbod, Bonadea, Bonnieturton, Boomshanka, Brenont, Brooser93, Brutaldeluxe, C777, Callmederek, Caltas, Camilo Sanchez, CanOfWorms, Capricorn42, CardinalDan, Carl Logan, Catterick, CharlesGillingham, Charlesdrakew, CharlieEchoTango, Chase me ladies, I'm the Cavalry, ChaseKiwi, Chblanchard, Chicen, Chienlit, Chobbs123, Chowbok, Chris jarvis, Chris the speller, ChrisRed, Chrisjj, Christian75, Ciroa, Civil Engineer III, Cnyborg, Cockles52, Colonies Chris, Conversion script, CoupleOfAvengers, Crococolana, Cyktsui, DJ Clayworth, DMZ77724, DMcMPO11AAUK, Dachshund, Dairyhouse, Dancingwombatsrule, Darkspots, Darwinek, David.Monniaux, DeadEyeArrow, Debz1983, Deciver132, Delicious carbuncle, Den fjättrade ankan, Denny Crane, DerHexer, Dergraaf, Dfrankow, Dimitrii, Dina, Discospinster, Divercol, Dividing, Doc glasgow, Docu, Dormskirk, Dpaajones, Dr. Blofeld, DrFrench, EdC, Eeekster, El C, Emijrp, Enchanter, Enviroboy, Eob, Eog1916, Epbr123, Ericoides, Esperanto, Eu.stefan, Everyking, Exir Kamalabadi, F1megastar, Fabartus, Favonian, Finavon, Flockmeal, Folks at 137, FrostySnows, Fsbr1908, Furrykef, Gaianauta, Gaius Cornelius, Gerry Lynch, Ghaly, Ghmyrtle, Ginar1, Gkedia mnnit, Godefroy, Good Olfactory, Goran.S2, Graham87, Gray62, GringoInChile, Gudeldar, Hadal, Halsteadk, Hardouin, HeartofaDog, Hedgehog, Hgrosser, HighKing, Hmains, Hmg210, Hugo999, Hunting dog, Iain99, IanDavies, Iantresman, Ibmab, Ilmari Karonen, Infrogmation, Iridescent, J.delanoy, JRPG, JaGa, JackofOz, JackyR, Jacobhenrytucker, Jacoplane, Jaho, Jake Spooky, Jamirocrazy, Janko, Jarle fagerheim, Jauhienij, Jayapura, Jennica, Jezzerk, Jgmccabe, Jimthedim, Jj137, Jmlk17, Jnestorius, JoanneB, Joe72, Johan Elisson, John Vandenberg, Jonzky, Jooler, Joowwww, Josemanimala, Joseph Solis in Australia, Joshurtree, Joyous!, Jrcla2, Jwillbur, Jza84, Kajmal, Kanags, Katsp8, Keith D, Kelborn, Kelisi, Kelvinc, Kember, Kingal86, Kiwipete, Klenod, Koplimek, Korg, Kostisl, Kumioko (renamed), Kurtle, Kurykh, Kwesiidun91, Lacrimosus, Laurencetimms, Lcawte, Leapsorymt, LeaveSleaves, Lemonade100, Leszek Jańczuk, Lightmouse, Like You Never Did See, LincsPaul, Little Mountain 5, Lord Cornwallis, LorenzoB, Louis R14, Lstanley1979, Luis wiki, Luna Santin, M-le-mot-dit, MBisanz, MJCdetroit, MOOGOLOPAZ, MPD01605, MWAK, MacRusgail, MacTire02, Magister Mathematicae, Maglev2, Mahahahaneapneap, Mais oui!, Malcolm Farmer, Malick78, Man vyi, Mandsford, Mani1, Marco79, Mariocki, Markussep, Marnanel, Marshman, Martin451, MartinBiely, Martynas Patasius, Maser Fletcher, Mattgirling, Mattsday, Mav, Maximaximax, Mboverload, Mcbloggart, Mclay1, Mdellapinna, Meco, Mhockey, Miaow Miaow, Michael J, MichaelTinkler, MidnightBlueMan, Miesianiacal, MikeF9, Mindsknife, Mintguy, Mjroots, Mnagaraja, Modest Genius, Mohsenkazempur, Montrealais, Morwen, Mr0t1633, MrGulli, Ms2ger, Mucky Duck, Naturenet, Naughtygirl69, Ndoak, Nealmueller, Neddyseagoon, Neelix, Neutrality, Nickrds09, Nightstallion, Nivix, Noel Streatfield, Noodleki, Noon, Notuncurious, Nthep, OMINOREG, Olivier, Pablomfa, Palica, Parhamr, Paul J Williams, Paulc206, Paxsimius, Pediac, Pharaoh of the Wizards, Piano non troppo, Pichpich, Pigsonthewing, Pippu d'Angelo, Pit, Plutonium27, Pne, PoleyDee, Pollinator, Quadell, Quentin X, Quest for Truth, Qwyrxian, R'n'B, Radagast3, Raghith, Ragib, Randavance, Rannpháirtí anaithnid, Raptus Regaliter Cattus Petasatus, Ricelovva, Rich Farmbrough, Richard Arthur Norton (1958-), Richard Harvey, Rjwilmsi, Rmhermen, Roofbird, Rossenglish, Roxi2, RxS, Ryanmcdaniel, SU Linguist, Samuel, Sango123, Sannse, Sfck1973, Sheridan, SkepticalRaptor, SkonesMickLoud, Skraz, Smartse, Smashville, Snigbrook, Snowdog, Socrates2008, Somerut, Sonikwlf, Sonjaaa, Soumyasch, SpookyMulder, Stan Shebs, Steinsky, SteveF48, Stupid Corn, Stymphal, Styrofoam1994, SunCreator, Swid, Tarquin, Tckma, Ted Longstaffe, Tempshill, Teratornis, Textorus, Thatperson, The Skywatcher and me, TheAllSeeingEye, TheHYPO, Thelb4, ThinkingTwice, Tholme, Thumperward, Tibor, Tide rolls, Timrollpickering, Timwi, Toby Bartels, Tom Meijer, Tomsparely, Tony Sidaway, Topdog1991, Trevor MacInnis, True Pagan Warrior, TwoOneTwo, UkPaolo, Ukexpat, Una Smith, Unschool, Urbanmyth139, Van helsing, Vanished user 03, Venatico, Vicki Rosenzweig, VolatileChemical, VraelVhalen, Vrenator, W0000t, Wavelength, Widefox, WikHead, William Avery, Wimt, WojPob, WolfgangFaber, Woohookitty, Xls, Xtreme racer, Yafarawla, Youandme, Zadradr, Zarius, Zazou, Zoe, Zoicon5, Zundark, Zzxc123, Зелёный Кошак, 614 anonymous edits

Mediterranean_Sea *Source*: http://en.wikipedia.org/w/index.php?title=Mediterranean_Sea *Contributors*: 2ko, A-giau, A.K.Khalifeh, A.L., Aarchiba, Abce2, Abemartin87, Absar, Absolutadam802, Acalamari, Acroterion, Adam Carr, Addshore, AdjustShift, Aggarank, Ahoerstemeier, Akanemoto, Akdenizg, Al-Andalus, Alansohn, AlgebraicSpore, Alnitak1971, Alsandro, AmRadioHed, Amalthea, Amarkov, Amazonien, Ancheta Wis, Andres, Andrwsc, Andy Dingley, Angelbo, Angr, Ani medjool, Anicic, Animum, Anna Frodesiak, Anthony Appleyard, Anxietycello, Arakunem, Arch dude, Artorius, Artreve, Artsandcraftsaremyfriends, Ascended Wraith, AshleyBird11, Ashleybird56, Ashleyisachild, Ashleyx123, Assaf050, Astavrou, Atgn148, Athaenara, Attilios, AuburnPilot, Aude, Aussieturk79, Autonova, Avatarion, B7582, BD2412, Babbage, Babelmed, BaboonOfTheYard, Backspace, BanyanTree, Bathrobe, Bazonka, Beaukarpo, Beivushtang, Ben-Zin, Bender235, Bentley4, Ber06122, Berkay0652, BhaiSaab, Big Adamsky, BigBossBlues, Bigtuurd12345789qwerty, BilboBaggans, Bill the Duck, Bill52270, BinAl-Turki, Bkell, Black Falcon, Blackfoxrunning, Blehfu, Bletch, Blog...Blog, BlueMoonlet, Bluemoose, Bobblewik, Bobbyhank1, Bobcats 23, Bobo192, Bogdangiusca, Bongwarrior, Bosonic dressing, BradBeattie, Brando130, Brian Pearson, Brighat, Brighterorange, British TV, Brutaldeluxe, Bryan Derksen, Bugoslav, By ram, C.Fred, CPColin, Camembert, Camoka4, Can't sleep, clown will eat me, Carlossuarez46, Carter5600, Causantin, Causteau, Chaser, Che829, Chris the speller, Chris.urs-o, Christian List, Chtito, Cjkporter, Ckatz, Cla68, ClockworkLunch, Cncs wikipedia, Cnilep, Cntras, Cocoma, Colchicum, Colonies Chris, CommonsDelinker, Conversion script, Cornischong, Corpx, Costa1143, Courthistorian, Cplakidas, Crnorizec, Crum375, Crusadeonilliteracy, Cserlajos, Csnewton, Css, Ctjf83, Cyanolinguophile, D6, DMacks, DOSGuy, Dan East, DandyDan2007, Daniel575, DanielCD, DarkFalls, Darth Mike, David Kernow, Dawn Bard, Dbo789, Dcsohl, Deanb, Deanjens, Deanos, Decltype, Dekimasu, Delaraha, Deli nk, Delirium, Demiurge, DerHexer, Dgw, Diannaa, Diimplezz, Dijxtra, Discospinster, Divanino, DiverDave, Djhguhuguiofighr, Djzare, Dleisawitz, Dmn, Docpi, Docu, Donarreiskoffer, Donreed, Dori, Dotdotlar, Doug, Doughmuffins, Dreadstar, Drhex, Drmaik, DubaiTerminator, Duja, DuncanPed, Dysepsion, Dysmorodrepanis, EL Willy, Earrings2max, Eboracum, EdBever, Edgar181, Edmundbaird, Eeekster, Egemens, Ehrenkater, Ekarfi13, Ekm02001, El C, Eleassar, Elmondo21st, Enchanter, Enea95, Enochlau, EoGuy, Eob, Eomermirimon, Epbr123, Eposse, Eskinat, Etams, Etz Haim, Eu.stefan, EugeneZelenko, Eupator, Excirial, Ezra Wax, FDV, FF2010, Fama Clamosa, Fastabbas, Fastily, FayssalF, Fifelfoo, Fireswordfight, Fishal, Flavius Belisarius, FloNight, Florian Blaschke, Fordsfords, Fourohfour, Francisco Valverde, Francisx, Friends147, Frymaster, Funandtrvl, Funnyhat, Fusionmix, Future Perfect at Sunrise, Gadfium, Gaius Cornelius, Galmicmi, Gary King, Garzo, Gavrilis, Gdr, General Eisenhower, GenestealerUK, Gensanders, GeoWriter, Geologyguy, George Al-Shami, Getoar TX, Gibmetal77, Gidip, Gilgamesh, Gilisa, Gilliam, Gioto, Gjd001, Glane23, Golbez, Graham87, Gramaic, Grandiose, GregU, Grundle2600, Grutness, Grylliade, Gtg827x, Gurch, Gwaz, Gwernol, Ha4iu5hj, Hadal, Hahahihihoho, Hairy Dude, Hans Dunkelberg, Harjk, Hashem sfarim, Haus, Haya1993, Hayabusa future, Hazerbey, Hectorian, HeeferFan, HeikoEvermann, HermaniscusVII, Heryu, HiDrNick, HighPriest15, Hike395, Historikos, Hmains, Hung pluto2003, HybridBoy, Hydrogen Iodide, II MusLiM HyBRiD II, IZAK, IanDavies, Ibizastorm, IceUnshattered, Idaltu, Igoldste, Imbris, Ingolfson, IrishAbroad, IstvanWolf, Itsmine, Ivirivi00, Ixfd64, J.delanoy, JDPhD, JForget, JFreeman, Jack69, Jacob.jose, Jagged 85, Jalwikip, Jamesontai, Jamie C, Jarble, Jatterb, Jaume87, Jayen466, Jb88gogg, Jcmolly, Jeancey, Jeffq, Jennavecia, Jeronimo, Jerycrimes, Jimtaip, Jmundo, Jntg4, JoeSmack, Jose77, Joy, Joyous!, Jpk, Jredmond, Juka, Jusdafax, JustPhil, Jvlm.123, KRBN, Kablammo, Kane5187, Karam.Anthony.K, Karen Johnson, Kayau, Kcowolf, Ketiltrout, Kevinalbania, Kevkevinjb, Khoikhoi, Kickorn, Kilhan, Killiondude, Kingpin13, Kirachinmoku, Kizildere, Kman543210, Kmusser, Knolls, Knutux, Koavf, KongminRegent, Kordas, Kristoferb, Kudret abi, Kungfuadam, Kupirijo, Kurieeto, Kurvanas, Kyle1278, LFaraone, LOLOLOLguy, Lacrimosus, Lambtron, LarryJeff, Lavateraguy, Lawrenceyuen, Lenticel, Liftarn, Lightmouse, Lilac Soul, Linaduliban, Lockesdonkey, Lord Hawk, Lradrama, Luanda e benfica, Luna Santin, Luwilt, M dorothy, MCintern, MEDIO, MER-C, MJCdetroit, MK8, Macedonian, Magioladitis, Magus732, Malinaccier, Malo, Man vyi, Marcbel, MarcoTolo, Markussep, Marquez, Martinezste, Materialscientist, MathKnight, Matt Crypto, Matt908, Maurice27, Mav, Maximaximax, Mazito, Mboverload, McSly, Mcorazao, Mediterano, Melih Yavuz, Menchi, Mente et malleo, Mentifisto, Merlinsorca, Mgiganteus1, Mhese, MiShogun, Micaburn, Michael Shields, Miguel303xm, Mikenorton, Mimmo46, Minimac, Mirv, Mitsver, Mllturro, Mmcannis, Momirt, Mooo, Moro 2088, Mothmolevna, Mountolive, Moverton, Mr Smiley, Mr. Triceratops, MrOllie, Mserena88, Msuave123, Mttll, MuZemike, Mulder1982, Mysdaao, N-k, N5iln, NTox, NYArtsnWords, Nakon, NatureA16, Nayvik, Nbuddoff, Nealmcb, Nebi, Necirsad, Neilc, NeroN BG, Neswtuop, Neutrality, NewEnglandYankee, Nick UA, Nita101, Nivek590, Nixeagle, Njaelkies Lea, Nk, Nol888, Nomenphile, NormanEinstein, Northamerica1000, Nubiatech, ONUnicorn, Odysses, Ohnoitsjamie, Okapi, Oleg Alexandrov, Olivier, Ollios, Ongar the World-Weary, Orges, Originalwana, OscarKosy, OwenBlacker, Oxymoron83, PAM - PI, Panairjdde, Parakalo, Pascal.Tesson, Passportguy, Patrick, Paul August, Paul Drye, Paul H., Pax85, Pennywisdom2099, Pepijn Koster, Percy Snoodle, Pere Serafi, Peregrine981, Perique des Palottes, Persian Poet Gal, Peter Greenwell, Peterlin, Pfly, Phil Bastian, Philip Trueman, Philly boy92, PhnomPencil, Piano non troppo, Piazzi, Picus viridis, Pietru, Pjamescowie, Planders, Plot Spoiler, Politis, Pollinator, Polylerus, Postdlf, Potes, Pourquoi Cherie, Pperos, ProudJewNo1, Proxima Centauri, Psychlopaedist, Ptolion, Purplefire, QuantumEleven, Qxz, R'n'B, RTucker, RaNdOm26, Ramirez72, Ratzer, Raven in Orbit, RavenRasputin, RaviC, Reaverdrop, RedCoat10, Redvers, Rennyhart, Res2216firestar, RetiredUser2, RexNL, Richdesign, Rizniz, Rje, Rjensen, Robin Hood 1212, Robomaeyhem, Romanzer, Ronline, Roylee, Rspits, SJP, SU Linguist, Saforrest, Sagaci, Sahilm, Saintmesmin, Sal73x, Sam Blacketer, Sardanaphalus, Scarian, SchwarzKatze, Scottmilligan, Selma Kaufmann, Shanoman, Shelblake, Shephia, Shizhao, Shoeofdeath, Sideshow Bob, SilentResident, Silver Edge, Silvestre, SimonP, Siswick, Sjakkalle, Skizzik, Slakr, Slowking Man, Sluzzelin, Snek01, Snigbrook, Snowmanradio, Snowolf, Some jerk on the Internet, Someguy1221, Somethingvacant, SovietCanuck, Sp33dyphil, SpeedyGonsales, SpookyMulder, Sprinter77, SpuriousQ, Stemonitis, Stephen G. Brown, Stepheng3, Stevertigo, Sting-fr, Storm Rider, Subtropical-man, Suffusion of Yellow, SunCreator, Supersexyspacemonkey, Supreme Deliciousness, Sweetxpea101, Symane, TAKASUGI Shinji, THEN WHO WAS PHONE?, TShilo12, Takeaway, Taki223, Tarquin, Tasc, Tbc2, Tbhotch, Techman224, Teemu Ruskeepää, Teles, That Guy, From That Show!, The Thing That Should Not Be, Theb2b, Themadgolfer, Theresa knott, Thijhgf, Thingg, Thorwald, Thryduulf, Tide rolls, Til Eulenspiegel, Timir2, TintininLisbon, Tobby72, Tom Radulovich, Tot-futbol, Tot12, Tpbradbury, Trevor MacInnis, Trud the Borg, Trusilver, Tushar858, Twest2665, Ubiq, Ultraviolet scissor flame, Umedard, Ungunga, Uriber, Useight, Valenciano, Valery Kulaga, Valley2city, Van helsing, Vancouverguy, Varana, Veinor, Versus22, Vice regent, Vicloic, Virinikimiki, Vishnava, Vneiomazza, Vsmith, WHATaintNOcountryIeverHEARDofDOtheySPEAKenglishINwhat, Waggers, Wapp, Wavelength, Waycool27, West.andrew.g, Wetman, WikiDao, WikiDusk5771, Wikigregor, WikipedianMarlith, William Avery, Wknight94, Wkolbe, Wolfkeeper, WolfmanSF, Woohookitty, Wuzzy, Ww2censor, Xeryus, Xezbeth, Xollob, Yamara, Yosraebeidy, Youssefsan, Yuber,

Yuje, Zenzee, ZeroFuzion, ZooFari, Zozo2kx, Zzuuzz, Александър, يامن, 1224 anonymous edits

Harland_and_Wolff *Source*: http://en.wikipedia.org/w/index.php?title=Harland_and_Wolff *Contributors*: Aaronjhill, Ad Orientem, Ahoerstemeier, Ahoghill, Ala.foum, Andyminicooper, Aquitania, Ardfern, Asfarer, Atlan, BD2412, Bacardimayne, Bellhalla, Benea, BigFatBuddha, Bigjimr, Bigpad, Bjmullan, Bmclaughlin9, Brad Rousse, Britmax, C0lmjl, Camw, Canterbury Tail, Chase me ladies, I'm the Cavalry, Chris the speller, ChrisGualtieri, Cimbalom, Cjrother, Conay, ConradPino, Cordless Larry, Cosal, Cp fan, DJ Clayworth, Dale Arnett, DanMS, Dashers, Davewild, DavidMcVeigh, Davidships, DePiep, Denniss, Doady, Docu, Dom0803, Dormskirk, DrFrench, Dreacos, Duk, EPO, Emoscopes, Eregli bob, Fallschirmjäger, Fennessy, Filastin, Finavon, Fredrik, Gboyers, Ghisguth, Gr1st, GraemeLeggett, Greenshed, Ground Zero, Gsl, Guliolopez, H1523702, HMCNR, Hairhorn, Hammersfan, Haus, HecRPD, Hibernian, Holon67, Hugo999, Ian Dunster, IrishPete, Irishguy, Jason Leake, Jatkins, Jcrct, Jimvanm, John, JohnOwens, Johnfos, Jonchapple, Jonnymccullagh, Joseph Solis in Australia, Joshbaumgartner, Keallu, Keithgreer, Kevin Murray, Keys767, Ksucasey13, LIU, Lapsed Pacifist, LastInLine44, Ldopa, Leandrod, Littlerobbergirl, LukeM212, M-le-mot-dit, MBK004, MC MasterChef, MCBastos, MKevH, MTL513, Malcolma, Mariepr, Mark83, MarkSutton, Marker10, MartaRibeiro, Maury Markowitz, Maxis ftw, MechBrowman, Mel Etitis, Michaelbarr123, Million Little Gods, Minimac, Mintguy, Mirage-project, Mjroots, Mooretwin, Mr Mulliner, Mrmorodo, Msa1701, Mu301nh, Mushroom, Neddyseagoon, Newport, Nicke L, Nlu, Nohat, Nohomers48, OfficeBoy, Olivier, Oly35, One Night In Hackney, Onebravemonkey, Palmiped, Patsw, Paul, Peter Clarke, Petri Krohn, Phind, Pterre, Pwt898, Qwfp, R'n'B, Rhollenton, Richard Arthur Norton (1958-), Rjwilmsi, Rlandmann, Rockysmile11, Ropers, SD6-Agent, Saberwyn, Saburny, SammyM, SchuminWeb, Sciurinæ, Setanta747 (locked), Skeetch, SoLando, Sorath, Stan Shebs, Stavros1, Steelartisan, Stompernz, Stringybark, Techtonic, Teratornis, The Madras, The jenan, The undertow, Thomas Blomberg, Tide rolls, Titanic1234567891011, Tom servo, TomB123, Versus22, Waitingforbrain, Wereon, Whitbycolin, Wolfyuk2007, Xn4, Yuriybrisk, 224 anonymous edits

Gross_tonnage *Source*: http://en.wikipedia.org/w/index.php?title=Gross_tonnage *Contributors*: Aldebaran, Borgx, DePiep, Dual Freq, EdC, ErikvanB, GregorB, Hasan ozbey, Haus, Jimp, Julesd, Lambiam, Matt314, Najro, Nutster, Orhanghazi, OverlordQ, Patrick, Rjwilmsi, Tomas e, Tupsumato, Yeokaiwei, 15 anonymous edits

Fishguard *Source*: http://en.wikipedia.org/w/index.php?title=Fishguard *Contributors*: *niCoLE*, Agathoclea, Ajho, Alansplodge, AlexJ, Ambleston, Amywinehouse22, Apparition11, Ari Rahikkala, Axeman89, Ba-KaDüÜ, Barek, Ben Ben, Ben craig, Bevo74, Bikerams, Biscuittin, CRKingston, Calum MacÙisdean, Charles RB, Cnyborg, Culturenut, Dalliance, DeadEyeArrow, Deb, Dr. Blofeld, Drachenfyre, DsquaredUK, Edwardx, Ehrenkater, Enaidmawr, Fdp, Frelke, FruitMonkey, GRAHAMUK, Gaius Cornelius, Geni, Geopersona, Ger-more, Gjenvick, GoingBatty, Gwalarn, Hairyjohn, Haniii, Hmains, Hogyn Lleol, Hu12, IPSOS, Indiasummer95, Ixfd64, JBellis, Jamesd, Jameswilson, Jeremy Bolwell, Jimmy3d0, Jivecat, John254, Jonny-newcastle, Jza84, Karenjc, LeilaniLad, LinguisticDemographer, Lorde555, Lumos3, MRSC, MacSteaphain, Maias, Man vyi, Marek69, Marnanel, MartinDK, Martpol, Mattcymru, Meister, Merlin60, MilfordBoy991, Misortie, Neddyseagoon, Nevilley, Nico29, NormDor, Onebravemonkey, Optimist on the run, Owain, Pauldack, Piano non troppo, Picapica, Pinethicket, Pit-yacker, Postdlf, Pwimageglow, Qwertyco, Richerman, Rjwilmsi, Rls, Saebjorn, Scottime, Skinsmoke, Skung, Sloman, SoLando, Someguy1221, T0m, Tassedethe, TheFurr, Thumperward, Tim A Lee, Tim!, Trident13, Trilobite, Ukeu, Vanished user 90345uifj983j4toi234k, Vany90, Velella, Versageek, Vibrating-goats, Voceditenore, WOSlinker, Waggers, Weisinger, Whpq, Wireless Keyboard, Woohookitty, Zzuuzz, 108 anonymous edits

Rosslare_Harbour *Source*: http://en.wikipedia.org/w/index.php?title=Rosslare_Harbour *Contributors*: A bit iffy, Ardfern, Boothy443, BrownHairedGirl, DaGr8one, Fennessy, Frelke, Gnomon, Hyacinth45, Isainmdom, Jvlm.123, MER-C, Peter Clarke, Picapica, Piercerugby, Rannpháirtí anaithnid, Rosser1954, Sarah777, Tabletop, Trevor MacInnis, Ww2censor, Zoney, 8 anonymous edits

Irish_Sea *Source*: http://en.wikipedia.org/w/index.php?title=Irish_Sea *Contributors*: 1297, 16@r, AdjustShift, AirdishStraus, Akanemoto, Alangley, Alien from brixton, An Muimhneach Machnamhach, Andrwsc, Andy Dingley, Anlace, Aotearoa, Arcturus, Asarlaí, Aughnadarragh, Avenue, BD2412, BIL, Backspace, Bagofrags, Balsa10, Bazonka, Beagel, Biruitorul, Bjankuloski06en, Bkonrad, BopSmart, BritishWatcher, Bryan Derksen, CALR, CR7, CRKingston, CZmarlin, CecilWard, Celtic Harper, Chewy200890, Chorusman, Chris55, Chrisjj, Clarino1, Closedmouth, Cradel, Craigbuck, Crispmuncher, D.de.loinsigh, Daarznieks, Dankarran, Darranc, Dave Bellamy, Davidleeis14, Doradus, Dougweller, Dr Greg, Drachenfyre, Drlnpstr, Dwalker19, EamonnPKeane, Ed g2s, Erebus555, Esobocinski, Eu.stefan, Excerpted31, Factotem, Feline1, Fennessy, Fiachra10003, Frokor, Gaius Cornelius, Gary King, Gene Nygaard, Ghmyrtle, Golbez, Good Olfactory, GoodDay, Graham87, Guliolopez, Haldraper, Halsteadk, Heavy joker, Hibernian, HighKing, Hmains, Hywel Ashkenazy, IanDavies, Inwind, JaGa, JamesAM, Jay-W, Jimmy Slade, Jlang, John, Jrobin08, Jvlm.123, Kathar, Kcordina, Lapsed Pacifist, LarryJeff, Lightmouse, Lipanook, LordAmeth, Lugnad, MacTire02, Man of wealth and taste, MapsMan, Mark83, Markreidyhp, Mattis, Mauls, Mejor Los Indios, Menchi, Miq, Mowsbury, MrChile, MusicInTheHouse, NatureA16, NeilN, Neutrality, Nk, OettingerCroat, Ohconfucius, Oleg Alexandrov, Only, Ori, Orphan Wiki, Owen4004, OwenBlacker, Parkjunwung, PatGallacher, Patrickneil, Peter Clarke, Pigman, Pigsonthewing, Pillow2011, Pinethicket, Preston North End Dan, Publunch, R'n'B, RTG, RainbowOfLight, Rannpháirtí anaithnid (old), Rapido, Razorflame, Red King, RedWolf, Redlentil, Renata, Reyk, Rich Farmbrough, RingtailedFox, Rjwilmsi, Robdurbar, Robert Skyhawk, Rockpocket, Rsvp123, Saga City, Sannse, Sarah777, Seabhcan, Shelblake, Shephia, SimonP, Slawojarek, Slgrandson, Snappy, Sowen, Spellmaster, Stabilo boss, Stemonitis, Stephen Turner, Stephenb, Stepheng3, Stevesmith1983, Stevew2022, Stevvvv4444, Sumatious, TFOWR, TGCP, Tangerines, TicketMan, Tigershrike, Tim!, Trilobite, Trusilver, Umbertoumm, Urhixidur, Uthican, Velella, VinceBowdren, Vssun, Waggers, Wgh001, Wiki01916, WikiDon, Wikipéire, Woohookitty, Ww2censor, YUL89YYZ, Yekrats, Zeitgeist, ZeroJanvier, 144 anonymous edits

Image Sources, Licenses and Contributors

File:RR Isla de Botafoc.jpg *Source*: http://en.wikipedia.org/w/index.php?title=File:RR_Isla_de_Botafoc.jpg *License*: unknown *Contributors*: User:William Avery

File:Flag of Cyprus.svg *Source*: http://en.wikipedia.org/w/index.php?title=File:Flag_of_Cyprus.svg *License*: unknown *Contributors*: User:Vzb83

File:Sealink br logo.svg *Source*: http://en.wikipedia.org/w/index.php?title=File:Sealink_br_logo.svg *License*: unknown *Contributors*: Philphos

Image:Gb~brail.png *Source*: http://en.wikipedia.org/w/index.php?title=File:Gb~brail.png *License*: unknown *Contributors*: User:Btline

File:Ventouris ferries logo.png *Source*: http://en.wikipedia.org/w/index.php?title=File:Ventouris_ferries_logo.png *License*: unknown *Contributors*: Philphos

File:Flag of Panama.svg *Source*: http://en.wikipedia.org/w/index.php?title=File:Flag_of_Panama.svg *License*: unknown *Contributors*: -xfi-, Addicted04, Alkari, Duduziq, Fadi the philologer, Fry1989, Hystrix, Klemen Kocjancic, Liftarn, Mattes, Nightstallion, Ninane, Pumbaa80, Reisio, Rfc1394, TFCforever, Thomas81, ThomasPusch, Zscout370, Ö, Фёдор Гусляров, 19 anonymous edits

File:EnglishChannel.jpg *Source*: http://en.wikipedia.org/w/index.php?title=File:EnglishChannel.jpg *License*: unknown *Contributors*: Badzil, Dbenzhuser, Muriel Gottrop, 3 anonymous edits

File:English Channel.jpg *Source*: http://en.wikipedia.org/w/index.php?title=File:English_Channel.jpg *License*: unknown *Contributors*: This is part of http://www.lib.utexas.edu/maps/europe/france_pol91.jpg; claimed to be from the CIA (see http://www.lib.utexas.edu/maps/france.html)

File:Carte de la Manche.png *Source*: http://en.wikipedia.org/w/index.php?title=File:Carte_de_la_Manche.png *License*: unknown *Contributors*: User:Bra

File:North sea languages 900.png *Source*: http://en.wikipedia.org/w/index.php?title=File:North_sea_languages_900.png *License*: unknown *Contributors*: User:Dbachmann

File:Hermitage St Helier Jersey.jpg *Source*: http://en.wikipedia.org/w/index.php?title=File:Hermitage_St_Helier_Jersey.jpg *License*: unknown *Contributors*: Original uploader was Man vyi at en.wikipedia

File:UK Radar1940.JPG *Source*: http://en.wikipedia.org/w/index.php?title=File:UK_Radar1940.JPG *License*: unknown *Contributors*: User:Nasiruddin

File:NormandyCourcelles2JM.jpg *Source*: http://en.wikipedia.org/w/index.php?title=File:NormandyCourcelles2JM.jpg *License*: unknown *Contributors*: Original uploader was John at en.wikipedia

File:German World War II tower Jersey.jpg *Source*: http://en.wikipedia.org/w/index.php?title=File:German_World_War_II_tower_Jersey.jpg *License*: unknown *Contributors*: AnRo0002, Danrok, Darkone, Deadstar, Man vyi, Responsible?, Ronaldino, Teofilo, Wojsyl, 1 anonymous edits

File:Saintmalo.jpg *Source*: http://en.wikipedia.org/w/index.php?title=File:Saintmalo.jpg *License*: unknown *Contributors*: Caesius, Man vyi, NaSH, Nataraja, Olivier2, Pymouss, TCY, 3 anonymous edits

File:Spinnaker Tower night.jpg *Source*: http://en.wikipedia.org/w/index.php?title=File:Spinnaker_Tower_night.jpg *License*: unknown *Contributors*: Cbuckley

File:Baie-du-Havre 14 07 2005.jpg *Source*: http://en.wikipedia.org/w/index.php?title=File:Baie-du-Havre_14_07_2005.jpg *License*: unknown *Contributors*: User:Urban

File:Norman dictionary 1779 Kelham.jpg *Source*: http://en.wikipedia.org/w/index.php?title=File:Norman_dictionary_1779_Kelham.jpg *License*: unknown *Contributors*: Man vyi

File:Battle of Lepanto 1571.jpg *Source*: http://en.wikipedia.org/w/index.php?title=File:Battle_of_Lepanto_1571.jpg *License*: unknown *Contributors*: anonymous

Image:Strait of gibraltar.jpg *Source*: http://en.wikipedia.org/w/index.php?title=File:Strait_of_gibraltar.jpg *License*: unknown *Contributors*: NASA

Image:Gallipoli peninsula from space.jpg *Source*: http://en.wikipedia.org/w/index.php?title=File:Gallipoli_peninsula_from_space.jpg *License*: unknown *Contributors*: Takeaway

File:MEDCURR.GIF *Source*: http://en.wikipedia.org/w/index.php?title=File:MEDCURR.GIF *License*: unknown *Contributors*: Original uploader was Valery Kulaga at en.wikipedia

File:Mediterranean Relief.jpg *Source*: http://en.wikipedia.org/w/index.php?title=File:Mediterranean_Relief.jpg *License*: unknown *Contributors*: Alfio, Dbenbenn, DieBuche, Fagairolles 34, Gwillhickers, Kintetsubuffalo, Look2See1, Ras67, Subtropical-man

File:Flag of Spain.svg *Source*: http://en.wikipedia.org/w/index.php?title=File:Flag_of_Spain.svg *License*: unknown *Contributors*: Anomie

File:Flag of France.svg *Source*: http://en.wikipedia.org/w/index.php?title=File:Flag_of_France.svg *License*: unknown *Contributors*: Anomie

File:Flag of Monaco.svg *Source*: http://en.wikipedia.org/w/index.php?title=File:Flag_of_Monaco.svg *License*: unknown *Contributors*: User:SKopp

File:Flag of Italy.svg *Source*: http://en.wikipedia.org/w/index.php?title=File:Flag_of_Italy.svg *License*: unknown *Contributors*: Anomie

File:Flag of Malta.svg *Source*: http://en.wikipedia.org/w/index.php?title=File:Flag_of_Malta.svg *License*: unknown *Contributors*: Alkari, Fry1989, Gabbe, Homo lupus, Klemen Kocjancic, Liftarn, Mattes, Meno25, Nightstallion, Peeperman, Pumbaa80, Ratatosk, Rodejong, Zscout370, 5 anonymous edits

File:Flag of Slovenia.svg *Source*: http://en.wikipedia.org/w/index.php?title=File:Flag_of_Slovenia.svg *License*: unknown *Contributors*: User:SKopp, User:Vzb83, User:Zscout370

File:Flag of Croatia.svg *Source*: http://en.wikipedia.org/w/index.php?title=File:Flag_of_Croatia.svg *License*: unknown *Contributors*: User:MaGa

File:Flag of Bosnia and Herzegovina.svg *Source*: http://en.wikipedia.org/w/index.php?title=File:Flag_of_Bosnia_and_Herzegovina.svg *License*: unknown *Contributors*: User:Kseferovic

File:Flag of Montenegro.svg *Source*: http://en.wikipedia.org/w/index.php?title=File:Flag_of_Montenegro.svg *License*: unknown *Contributors*: User:B1mbo, User:Froztbyte

File:Flag of Albania.svg *Source*: http://en.wikipedia.org/w/index.php?title=File:Flag_of_Albania.svg *License*: unknown *Contributors*: User:Dbenbenn

File:Flag of Greece.svg *Source*: http://en.wikipedia.org/w/index.php?title=File:Flag_of_Greece.svg *License*: unknown *Contributors*: (of code) (talk)

File:Flag of Turkey.svg *Source*: http://en.wikipedia.org/w/index.php?title=File:Flag_of_Turkey.svg *License*: unknown *Contributors*: User:Dbenbenn

File:Flag of Syria.svg *Source*: http://en.wikipedia.org/w/index.php?title=File:Flag_of_Syria.svg *License*: unknown *Contributors*: see below

File:Flag of Lebanon.svg *Source*: http://en.wikipedia.org/w/index.php?title=File:Flag_of_Lebanon.svg *License*: unknown *Contributors*: Traced based on the CIA World Factbook with some modification done to the colours based on information at Vexilla mundi.

File:Flag of Israel.svg *Source*: http://en.wikipedia.org/w/index.php?title=File:Flag_of_Israel.svg *License*: unknown *Contributors*: "The Provisional Council of State Proclamation of the Flag of the State of Israel" of 25 Tishrei 5709 (28 October 1948) provides the official specification for the design of the Israeli flag. The color of the Magen David and the stripes of the Israeli flag is not precisely specified by the above legislation. The color depicted in the current version of the image is typical of flags used in Israel today, although individual flags can and do vary. The flag legislation officially specifies dimensions of 220 cm × 160 cm. However, the sizes of actual flags vary (although the aspect ratio is usually retained).

File:Flag of Egypt.svg *Source*: http://en.wikipedia.org/w/index.php?title=File:Flag_of_Egypt.svg *License*: unknown *Contributors*: Open Clip Art

File:Flag of Libya.svg *Source*: http://en.wikipedia.org/w/index.php?title=File:Flag_of_Libya.svg *License*: unknown *Contributors*: Various

File:Flag of Tunisia.svg *Source*: http://en.wikipedia.org/w/index.php?title=File:Flag_of_Tunisia.svg *License*: unknown *Contributors*: Alkari, AnonMoos, Avala, Bender235, Bradayy, Duduziq, Elina2308, Emmanuel.boutet, Flad, Fry1989, Gabbe, Juiced lemon, Klemen Kocjancic, Mattes, Meno25, Moumou82, Myself488, Neq00, Nightstallion, Reisio, Str4nd, TFCforever, Ö, Фёдор Гусляров, 9 anonymous edits

File:Flag of Algeria.svg *Source*: http://en.wikipedia.org/w/index.php?title=File:Flag_of_Algeria.svg *License*: unknown *Contributors*: User:SKopp

File:Flag of Morocco.svg *Source*: http://en.wikipedia.org/w/index.php?title=File:Flag_of_Morocco.svg *License*: unknown *Contributors*: User:Denelson83, User:Zscout370

File:Flag of Gibraltar.svg *Source*: http://en.wikipedia.org/w/index.php?title=File:Flag_of_Gibraltar.svg *License*: unknown *Contributors*: Alkari, Dbenbenn, Denelson83, Fry1989, Mattes, Nightstallion, Pumbaa80, 3 anonymous edits

File:Flag Ceuta.svg *Source*: http://en.wikipedia.org/w/index.php?title=File:Flag_Ceuta.svg *License*: unknown *Contributors*: User:Ulaidh

File:Flag Melilla.svg *Source*: http://en.wikipedia.org/w/index.php?title=File:Flag_Melilla.svg *License*: unknown *Contributors*: User:Ulaidh

File:Flag of the United Kingdom.svg *Source*: http://en.wikipedia.org/w/index.php?title=File:Flag_of_the_United_Kingdom.svg *License*: unknown *Contributors*: Anomie, Good Olfactory, Mifter

File:Flag of the Turkish Republic of Northern Cyprus.svg *Source*: http://en.wikipedia.org/w/index.php?title=File:Flag_of_the_Turkish_Republic_of_Northern_Cyprus.svg *License*: unknown *Contributors*: Absar, Alkari, Anime Addict AA, Artem Karimov, Bjs, Carnildo, Dbenbenn, FSII, Fry1989, Himasaram, Homo lupus, Juiced lemon, Mattes, Multichill, Neq00, Ninane, Nokka, Pumbaa80, Reuvenk, Slomox, Wutsje, 10 anonymous edits

File:Flag of Palestine.svg *Source*: http://en.wikipedia.org/w/index.php?title=File:Flag_of_Palestine.svg *License*: unknown *Contributors*: User:Orionist, user:Makaristos

Image:Bucht & Straße von Gibraltar.jpg *Source*: http://en.wikipedia.org/w/index.php?title=File:Bucht_&_Straße_von_Gibraltar.jpg *License*: unknown *Contributors*: User:Andinem

File:Torre della Pelosa.jpg *Source*: http://en.wikipedia.org/w/index.php?title=File:Torre_della_Pelosa.jpg *License*: unknown *Contributors*: User:Antofa

Image:Beirut-Sannine.jpg *Source*: http://en.wikipedia.org/w/index.php?title=File:Beirut-Sannine.jpg *License*: unknown *Contributors*: BlingBling10, Hailey C. Shannon, Linaduliban

File:Tunisia - Sicily - South Italy.jpg *Source*: http://en.wikipedia.org/w/index.php?title=File:Tunisia_-_Sicily_-_South_Italy.jpg *License*: unknown *Contributors*: User:DrFO.Jr.Tn

File:Flag of Sicily (revised).svg *Source*: http://en.wikipedia.org/w/index.php?title=File:Flag_of_Sicily_(revised).svg *License*: unknown *Contributors*: User:Angelo.romano

File:Flag of Sardinia.svg *Source*: http://en.wikipedia.org/w/index.php?title=File:Flag_of_Sardinia.svg *License*: unknown *Contributors*: User:Angelo.romano

File:Flag of Corsica.svg *Source*: http://en.wikipedia.org/w/index.php?title=File:Flag_of_Corsica.svg *License*: unknown *Contributors*: user:Patricia.fidi

File:Flag of Mallorca.svg *Source*: http://en.wikipedia.org/w/index.php?title=File:Flag_of_Mallorca.svg *License*: unknown *Contributors*: User:Himasaram

File:Port Autonome de Marseille.JPG *Source*: http://en.wikipedia.org/w/index.php?title=File:Port_Autonome_de_Marseille.JPG *License*: unknown *Contributors*: Vincent

File:Vrulja kod Omiša.jpg *Source*: http://en.wikipedia.org/w/index.php?title=File:Vrulja_kod_Omiša.jpg *License*: unknown *Contributors*: User:Ante Perkovic

Image:Etapa1.jpg *Source*: http://en.wikipedia.org/w/index.php?title=File:Etapa1.jpg *License*: unknown *Contributors*: User:Paubahi

Image:Etapa2.jpg *Source*: http://en.wikipedia.org/w/index.php?title=File:Etapa2.jpg *License*: unknown *Contributors*: User:Paubahi

Image:Etapa3muda.jpg *Source*: http://en.wikipedia.org/w/index.php?title=File:Etapa3muda.jpg *License*: unknown *Contributors*: User:Paubahi

Image:Etapa4muda.jpg *Source*: http://en.wikipedia.org/w/index.php?title=File:Etapa4muda.jpg *License*: unknown *Contributors*: User:Paubahi

File:Himantura uarnak egypt.jpg *Source*: http://en.wikipedia.org/w/index.php?title=File:Himantura_uarnak_egypt.jpg *License*: unknown *Contributors*: Aurimas Mikalauskas from Paliūniškis, Lithuania

File:Barceloneta 2007.jpg *Source*: http://en.wikipedia.org/w/index.php?title=File:Barceloneta_2007.jpg *License*: unknown *Contributors*: User:Otto Normalverbraucher

File:Aquaculture Western Greece 2004.jpg *Source*: http://en.wikipedia.org/w/index.php?title=File:Aquaculture_Western_Greece_2004.jpg *License*: unknown *Contributors*: User:AlMare

File:ForbysIbizaTown 02.jpg *Source*: http://en.wikipedia.org/w/index.php?title=File:ForbysIbizaTown_02.jpg *License*: unknown *Contributors*: Factumquintus, Miaow Miaow

File:Plage-de-la-courtade.jpg *Source*: http://en.wikipedia.org/w/index.php?title=File:Plage-de-la-courtade.jpg *License*: unknown *Contributors*: Baptiste Rossi

File:Cala Goloritze o.jpg *Source*: http://en.wikipedia.org/w/index.php?title=File:Cala_Goloritze_o.jpg *License*: unknown *Contributors*: delaere

File:Malta Coast.jpg *Source*: http://en.wikipedia.org/w/index.php?title=File:Malta_Coast.jpg *License*: unknown *Contributors*: User:Kristoferb

File:Cavtat Croatia 2008-10-07.JPG *Source*: http://en.wikipedia.org/w/index.php?title=File:Cavtat_Croatia_2008-10-07.JPG *License*: unknown *Contributors*: User:Ketone16

File:svetistefan1756.JPG *Source*: http://en.wikipedia.org/w/index.php?title=File:Svetistefan1756.JPG *License*: unknown *Contributors*: User:Einer flog zu Weit

File:Ksamil Beach.jpg *Source*: http://en.wikipedia.org/w/index.php?title=File:Ksamil_Beach.jpg *License*: unknown *Contributors*: Original uploader was IllyricumSacrum at en.wikipedia

File:Ναυάγιο Ζακύνθου.jpg *Source*: http://en.wikipedia.org/w/index.php?title=File:Ναυάγιο_Ζακύνθου.jpg *License*: unknown *Contributors*: User:Alexignatiou

File:Burjeslam.jpg *Source*: http://en.wikipedia.org/w/index.php?title=File:Burjeslam.jpg *License*: unknown *Contributors*: User:Ahmadac

File:BeirutRaouche1.jpg *Source*: http://en.wikipedia.org/w/index.php?title=File:BeirutRaouche1.jpg *License*: unknown *Contributors*: Eusebius, Gilgameshkun, Heretiq

File:MediterraneanSeaFromIsrael.jpg *Source*: http://en.wikipedia.org/w/index.php?title=File:MediterraneanSeaFromIsrael.jpg *License*: unknown *Contributors*: Matanya (usurped), Pruneau, Tangopaso

File:Coast of Alexandria, A view From Bibliotheca Alexandrina, Egypt.jpg *Source*: http://en.wikipedia.org/w/index.php?title=File:Coast_of_Alexandria,_A_view_From_Bibliotheca_Alexandrina,_Egypt.jpg *License*: unknown *Contributors*: User:Moro 2088

File:Beach benghazi 3.JPG *Source*: http://en.wikipedia.org/w/index.php?title=File:Beach_benghazi_3.JPG *License*: unknown *Contributors*: User:Dennixo

File:Hammametgolf.jpg *Source*: http://en.wikipedia.org/w/index.php?title=File:Hammametgolf.jpg *License*: unknown *Contributors*: Jaume Ollé

File:Les Aiguades.jpg *Source*: http://en.wikipedia.org/w/index.php?title=File:Les_Aiguades.jpg *License*: unknown *Contributors*: User:Vermondo

File:EL Jebha1.jpg *Source*: http://en.wikipedia.org/w/index.php?title=File:EL_Jebha1.jpg *License*: unknown *Contributors*: Original uploader was Jomny at en.wikipedia

File:Gibraltar-Europa-Point-LH-from-the-sea.jpg *Source*: http://en.wikipedia.org/w/index.php?title=File:Gibraltar-Europa-Point-LH-from-the-sea.jpg *License*: unknown *Contributors*: Paul Handley

Image:Harlandandwolff.png *Source*: http://en.wikipedia.org/w/index.php?title=File:Harlandandwolff.png *License*: unknown *Contributors*: User:Keithgreer

Image:Knocking off at Harland & Wolff, Belfast.jpg *Source*: http://en.wikipedia.org/w/index.php?title=File:Knocking_off_at_Harland_&_Wolff,_Belfast.jpg *License*: unknown *Contributors*: Original uploader was Stavros1 at en.wikipedia

Image:Edward James Harland Belfast.jpg *Source*: http://en.wikipedia.org/w/index.php?title=File:Edward_James_Harland_Belfast.jpg *License*: unknown *Contributors*: Original uploader was Peter Clarke at en.wikipedia

File:Draughting Office Harland & Wolff, Belfast by Karl Beutel 2009.jpg *Source*: http://en.wikipedia.org/w/index.php?title=File:Draughting_Office_Harland_&_Wolff,_Belfast_by_Karl_Beutel_2009.jpg *License*: unknown *Contributors*: User:Teufelbeutel

Image:Shipbuilding, Liverpool 1944.jpg *Source*: http://en.wikipedia.org/w/index.php?title=File:Shipbuilding,_Liverpool_1944.jpg *License*: unknown *Contributors*: FSII, Irate, Man vyi, SoLando

Image:H&W Cranes2.jpg *Source*: http://en.wikipedia.org/w/index.php?title=File:H&W_Cranes2.jpg *License*: unknown *Contributors*: Plastic Jesus (Dave)

File:'Resolution' in Belfast - geograph.org.uk - 636807.jpg *Source*: http://en.wikipedia.org/w/index.php?title=File:'Resolution'_in_Belfast_-_geograph.org.uk_-_636807.jpg *License*: unknown *Contributors*: Ardfern, Docu, Tetris L

File:'Isle of Inishmore' and 'Jonathan Swift' in Belfast - geograph.org.uk - 653773.jpg *Source*: http://en.wikipedia.org/w/index.php?title=File:'Isle_of_Inishmore'_and_'Jonathan_Swift'_in_Belfast_-_geograph.org.uk_-_653773.jpg *License*: unknown *Contributors*: Ardfern, Docu

Image:Ss Venetian Belfast.jpg *Source*: http://en.wikipedia.org/w/index.php?title=File:Ss_Venetian_Belfast.jpg *License*: unknown *Contributors*: Original uploader was Peter Clarke at en.wikipedia

Image:Naval Ensign of the United Kingdom.svg *Source*: http://en.wikipedia.org/w/index.php?title=File:Naval_Ensign_of_the_United_Kingdom.svg *License*: unknown *Contributors*: User:Pumbaa80

Image:Civil Ensign of the United Kingdom.svg *Source*: http://en.wikipedia.org/w/index.php?title=File:Civil_Ensign_of_the_United_Kingdom.svg *License*: unknown *Contributors*: User:Pumbaa80

File:Lod Schema.png *Source*: http://en.wikipedia.org/w/index.php?title=File:Lod_Schema.png *License*: unknown *Contributors*: Robert Blazek, Pezinok, Slovakia. I am the author and I grant it wikipedia for free use.

File:Fishguard Harbour.JPG *Source*: http://en.wikipedia.org/w/index.php?title=File:Fishguard_Harbour.JPG *License*: unknown *Contributors*: Original uploader was Tim A Lee at en.wikipedia

file:Pembrokeshire UK location map.svg *Source*: http://en.wikipedia.org/w/index.php?title=File:Pembrokeshire_UK_location_map.svg *License*: unknown *Contributors*: User:Nilfanion

File:Red pog.svg *Source*: http://en.wikipedia.org/w/index.php?title=File:Red_pog.svg *License*: unknown *Contributors*: Anomie

File:Fishguard pembrokeshire map.svg *Source*: http://en.wikipedia.org/w/index.php?title=File:Fishguard_pembrokeshire_map.svg *License*: unknown *Contributors*: Alexander Jones

File:Fishguard & Rosslare Railways and Harbours Company Cerificate.jpg *Source*: http://en.wikipedia.org/w/index.php?title=File:Fishguard_&_Rosslare_Railways_and_Harbours_Company_Cerificate.jpg *License*: unknown *Contributors*: F&RH&R co. Receipt

Image:Roslare Europort 02.jpg *Source*: http://en.wikipedia.org/w/index.php?title=File:Roslare_Europort_02.jpg *License*: unknown *Contributors*: Ardfern, FalseXflag, Gavigan

File:Locatie Ierse Zee.PNG *Source*: http://en.wikipedia.org/w/index.php?title=File:Locatie_Ierse_Zee.PNG *License*: unknown *Contributors*: edited by M.Minderhoud

File:IMG Brittas Bay1529.jpg *Source*: http://en.wikipedia.org/w/index.php?title=File:IMG_Brittas_Bay1529.jpg *License*: unknown *Contributors*:

File:IMG DublinBay1467a.jpg *Source*: http://en.wikipedia.org/w/index.php?title=File:IMG_DublinBay1467a.jpg *License*: unknown *Contributors*: Docu

File:Barrow Offshore wind turbines.jpg *Source*: http://en.wikipedia.org/w/index.php?title=File:Barrow_Offshore_wind_turbines.jpg *License*: unknown *Contributors*: User:Andy Dingley

MIX
Papier aus verantwortungsvollen Quellen
Paper from responsible sources
FSC® C105338

Printed by Books on Demand GmbH, Norderstedt / Germany